Coatings for Cultural Heritage: Cleaning, Protection and Restoration

Coatings for Cultural Heritage: Cleaning, Protection and Restoration

Guest Editor

Yumin Du

MDPI

Basel • Beijing • Wuhan • Barcelona • Belgrade • Novi Sad • Cluj • Manchester

Guest Editor
Yumin Du
Institute of Culture and
Heritage
Northwestern Polytechnical
University
Xi'an
China

Editorial Office
MDPI AG
Grosspeteranlage 5
4052 Basel, Switzerland

This is a reprint of the Special Issue, published open access by the journal *Coatings* (ISSN 2079-6412), freely accessible at: https://www.mdpi.com/journal/coatings/special_issues/cultural_heritage_cleaning_protection_restoration.

For citation purposes, cite each article independently as indicated on the article page online and as indicated below:

Lastname, A.A.; Lastname, B.B. Article Title. *Journal Name* **Year**, *Volume Number*, Page Range.

ISBN 978-3-7258-5737-1 (Hbk)
ISBN 978-3-7258-5738-8 (PDF)
https://doi.org/10.3390/books978-3-7258-5738-8

Contents

About the Editor . . . . **vii**

Weihan Zou and Sok-Yee Yeo
Investigation on the Painting Materials and Profile Structures Used in Ancient Chinese Folk Architectural Paintings by Multiple Analytical Methods
Reprinted from: *Coatings* **2022**, *12*, 320, https://doi.org/10.3390/coatings12030320 . . . . **1**

Na Su, Bo Yang, Wenwu Chen, Linrong Xu and Yongwei Li
Influence of Prevailing Wind Direction on Sapping Quantity of Rammed Earth Great Wall of the Ming Dynasty
Reprinted from: *Coatings* **2022**, *12*, 707, https://doi.org/10.3390/coatings12050707 . . . . **15**

Qiangqiang Pei, Bo Zhang, Dongjuan Shang, Qinglin Guo, Jinjing Huang and Jing Zhu
Characteristics of Temperature Field of Rammed Earth Wall in Arid Environment
Reprinted from: *Coatings* **2022**, *12*, 735, https://doi.org/10.3390/coatings12060735 . . . . **26**

Xiangpeng Yu, Kai Cui, Xue Hu, Qian Qiao and Guojun Zhang
Properties and Compatibility Assessment of the Slurry with Quicklime for Fissure Restoration of Earthen Sites in Northwest China
Reprinted from: *Coatings* **2022**, *12*, 1255, https://doi.org/10.3390/coatings12091255 . . . . **53**

Zhenzhen Ma, Lu Yang, Liqin Wang, Václav Pitthard, Tatjana Bayerova, Gabriela Krist and Xichen Zhao
The Influence of Natural Aging Exerting on the Stability of Some Proteinaceous Binding Media Commonly Used in Painted Artworks
Reprinted from: *Coatings* **2022**, *12*, 1522, https://doi.org/10.3390/coatings12101522 . . . . **73**

Guopeng Wu, Kai Cui, Chengrui Ge and Xiangpeng Yu
The Process and Mechanisms of Freezing and Thawing Deterioration of Petroglyphs under Snowmelt and Rainfall Infiltration
Reprinted from: *Coatings* **2023**, *13*, 543, https://doi.org/10.3390/coatings13030543 . . . . **85**

Dan Liu, Xiaolong Tian, Dong Zhang, Xianjing Zhou, Nana Li and Yajun Zhao
Scientific Research on a Gold- and Silver-Inlaid Bronze Zun from the Han Dynasty
Reprinted from: *Coatings* **2023**, *13*, 1480, https://doi.org/10.3390/coatings13091480 . . . . **99**

Xing Zhao, Xia Li, Siyu Zhang, Qing Niu, Zongmin Li and Cheng Xue
Investigation of Whitening Mechanism on Cultural Relic Surfaces Treated with Paraloid B72
Reprinted from: *Coatings* **2024**, *14*, 1240, https://doi.org/10.3390/coatings14101240 . . . . **109**

Yupeng Li, Sokyee Yeo, Weihan Zou and Shibing Dai
Assessing the Fire Properties of Various Surface Treatments on Timber Components in Ancient Chinese Buildings: A Case Study from the Xianqing Temple in Changzhi, Shanxi, China
Reprinted from: *Coatings* **2024**, *14*, 1326, https://doi.org/10.3390/coatings14101326 . . . . **123**

Federica Antonelli, Sara Iovine, Carlotta Sacco Perasso, Natalia Macro, Eleonora Gioventù, Filippo Edoardo Capasso and Marco Bartolini
Essential Oils and Essential Oil-Based Products Compared to Chemical Biocides Against Microbial Patinas on Stone Cultural Heritage
Reprinted from: *Coatings* **2024**, *14*, 1546, https://doi.org/10.3390/coatings14121546 . . . . **139**

Georgios Tsairis, Athina Georgia Alexopoulou, Nicolaos Zacharias and Ioanna Kakoulli
Close-Range Photogrammetry and RTI for 2.5D Documentation of Painted Surfaces: A Tiryns Mural Case Study
Reprinted from: *Coatings* **2025**, *15*, 388, https://doi.org/10.3390/coatings15040388 . . . . **154**

About the Editor

Yumin Du

Yumin Du is a tenured associate professor at the Institute of Culture and Heritage, Northwestern Polytechnical University, and a master's supervisor. He has been engaged in research on the protection of earthen sites for a long time. He has presided over four national and provincial-level projects, including the National Natural Science Foundation, the Shaanxi Social Science Foundation, the Shaanxi Two-Chain Integration Key Special Project: Sub-Project of Major Special Project for Cultural Heritage Protection and Utilization, and the China Postdoctoral Science Foundation. He has also led one open project of the key laboratory of the Ministry of Education and four university-level projects. In recent years, he has published over 20 high-level papers in authoritative domestic and international journals in the fields of cultural relic protection and archeology. He serves as the deputy secretary-general of the Shaanxi Rock and Soil Mechanics and Engineering Society, the secretary-general of the Rock and Soil Cultural Relics Protection Special Committee, and a member of the Unsaturated Soil and Special Soil Discipline Group of the Geotechnical Engineering Branch of the Chinese Civil Engineering Society. He is also a reviewer for more than ten top journals in the fields of cultural heritage preservation and geotechnical engineering, such as *Journal of Cultural Heritage*, *Acta Geotechnica*, *Geomorphology*, etc.

coatings

MDPI

Article

Investigation on the Painting Materials and Profile Structures Used in Ancient Chinese Folk Architectural Paintings by Multiple Analytical Methods

Weihan Zou and Sok-Yee Yeo *

Department of Architecture, School of Human Settlements and Civil Engineering, Xi'an Jiaotong University, Xi'an 710049, China; zouweihan78@stu.xjtu.edu.cn
* Correspondence: yeosokyee@mail.xjtu.edu.cn

Abstract: In order to understand the painting materials and profile structures used in the creation of Chinese ancient folk architectural paintings, the architectural paintings of the Bell and Drum Tower at Fushan Temple in Shaanxi Province of China was investigated. In this study, optical microscopy, Fourier transform infrared spectroscopy (FTIR), micro-Raman spectroscopy, scanning electron microscopy-energy-dispersive X-ray spectroscopy (SEM-EDS), and chemical experimental methods were used. The profile structures, and the elemental and chemical compositions of the pigments and binders in the ground layer of the 12 samples were determined. Results showed that the painting profile structures in both towers comprised of pigment layer, underpainting layer and ground layer, where no starch was found in the chalk ground layer. The pigment layer mainly comprised of iron oxide red, chalk, lapis lazuli, carbon black, green earth, and a dark green pigment that is formed by a combination of carbon black and green earth. To-date, the use of green earth pigment in Chinese architectural paintings was not found in Chinese related academic literature; hence, this study marked the first identification of green earth being used in the architectural paintings. Tung oil, commonly used in Chinese architectural paintings as a binder for pigment, was also identified in the samples. Results from this study will serve as an important reference for better scientific investigation methods on ancient Chinese folk painting materials.

Keywords: ancient folk architectural paintings; pigments; binders; profile structures; FTIR spectroscopy; micro-Raman spectroscopy; SEM-EDS

1. Introduction

Ancient Chinese architectural paintings, commonly used to decorate the surfaces of building columns, beams, walls and other structural elements, consist mainly of pigment and ground layer. These paintings are important features of ancient Chinese architecture, with great historical, cultural, and artistic value [1]. However, due to natural degradation over time, the paintings on the building faced deterioration problems such as flaking, peeling, and fading of the pigment layer and ground layer, thereby leading to the gradual loss of cultural and artistic heritage [2]. Many types of ancient Chinese architectural paintings are still in existence and are widely distributed, but to-date, limited studies can be found on the proper evaluation methodology for the original production processes of ancient Chinese architectural painting. In particular, previous studies mostly focused on the study of ancient Chinese official architectural paintings, such as the Cining Palace paintings [3], the Xi'an Drum Tower paintings [2], and the Foguang Temple paintings [4]. The study of ancient folk architectural paintings, on the other hand, has rarely been discussed. Thus, there is an urgent need to understand the painting profile and material composition before conservation works can be carried out properly. The Shaanxi Fushan temple is, by far, one of the few remaining folk temples in China that was first established in the Tang dynasty and has witnessed thousands of years of ancient Chinese civilization. Hence, to better understand the painting profile structures and materials used in the production of the ancient Chinese folk architectural paintings, the architectural paintings of the Fushan Temple was selected as a representative case due to its long historical background.

Located in Heyang county of Shaanxi Province, the ancient architectural complex of Fushan temple is the fourth batch of cultural relics protection unit listed in the Shaanxi Province (Figure 1). Based on historical records, the building complex was first established in the Tang dynasty, and had undergone several major constructions between the Ming and Qing dynasties. Last recorded repair and construction works, involving the existing Bell and Drum Towers and the additions of new ancillary structures (Sanqing pavilion, Praying pavilion, Earth pavilion and Wall of Blessing) took place in 1882 AD. Due to natural weathering over time, the architectural paintings on the Drum and Bell Towers faced severe deterioration problems, such as flaking, peeling, and discoloration, subsequently leading to exposure and gradual decay of the wooden components. As the Bell and Drum Towers are the last remaining group of original buildings left within the complex, the decision was made to select samples with the most severe damage condition from different parts of the Bell and Drum Tower for investigation and analysis.

Figure 1. (**a**) the location of Fushan Temple in Weinan, Shaanxi Province, China; (**b**) the entrance of Fushan Temple; (c) the Drum tower; (**d**) the Bell tower.

In recent years, micro-Raman spectroscopy, Fourier transform infrared spectroscopy (FTIR), scanning electron microscopy-energy-dispersive X-ray spectroscopy (SEM-EDS), and other analytical techniques have been used widely to investigate the composition, morphology, and microstructure of various painted artworks [5–8]. Raman spectroscopy has been used to analyze a wide range of pigment compositions in artwork because it is non-destructive and can potentially characterize mineral materials [9,10]. For example, Proietti et al. [11] used Raman spectroscopy to identify the composition and pigments in Nubian mural painting layer, and confirmed the presence of pigments such as hematite, gypsum, and atacamite. The materials of mural paintings from Liao dynasty in Xuanhua, China, were also identified by micro-Raman spectroscopy and indicated that the pigments of plaster, hematite ultramarine, and carbon were applied [6].

Similar to micro-Raman spectroscopy, FTIR spectroscopy can be used for qualitative analyses of artwork materials and since it complements the results obtained by Raman spectroscopy, FTIR spectroscopy is also considered as an important method for the detection of organic materials with polar molecules [12,13]. Cheilakou et al. [7], for example, identified the presence of organic binders in various types of pigments in addition to cinnabar, ochre, and celadonite, used in the Byzantine wall paintings by Fourier transform infrared spectroscopy, and confirmed that the binders were protein-based substances by analyzing their infrared spectral absorption peaks. Moretti et al. [12] also used FTIR spectroscopy to determine the composition of Argentinean murals and confirmed the presence of gypsum and drying oil.

SEM-EDS and optical microscopy are particularly useful for observing the microscopic morphology and structure of painted artwork materials [14,15]. For example, in the case of the fresco sections in an Andean church, Tomasini et al. [16] determined the morphological characteristics of the layers and the elemental composition using optical microscopy and SEM-EDS, and confirmed that the sequence of the fresco's sections began with the pigment layer lining the surface, followed by white ground layer applied above the adobe wall. In the study conducted by Demir et al. [14], the painting techniques were also analyzed by optical microscopy and SEM-EDS, and the results indicated that the paintings were composed of a fine plaster and a thin paint layer over the plaster layer.

In this paper, profile structures of the samples from various parts of the Drum Tower and Bell Tower were analyzed using optical microscopy. Elemental and chemical compositions of the pigments, binders, and ground layer in the paintings were also qualitatively analyzed using micro-Raman spectroscopy, FTIR spectroscopy, and EDS. As noted in two important ancient Chinese building construction manuals, namely the Song dynasty "*Yingzao fashi*" [17] and the Qing dynasty "*Gongcheng Zuofa Zeli*" [18], starch (such as flour) was commonly used as one of the ingredients for the ground layer of architectural paintings. Hence, iodine staining tests were applied to the samples to determine whether starch was present in the ground layer of the paintings. Finally, based on the results obtained from the above tests, scientific methods on the selection of materials and techniques for the future restoration of ancient Chinese architectural paintings were proposed.

2. Methods and Experimental Design

2.1. Samples and Dyeing Experiments

Twelve samples were carefully extracted from the severely damaged portions of the Bell and Drum Towers' columns, beams and Queti (a beam-supporting brace structure). Based on the "Principles for the Conservation of Heritage Sites in China" [19], tiny fragments from the painting layers with severe deterioration or flaking condition were gently scrapped with a scalpel so as to reduce the damage done towards the historic decorations. The fragments were then placed in sample bags and sealed for subsequent laboratory tests and analysis. Details of the 12 samples are listed in Table 1 and Figure 2.

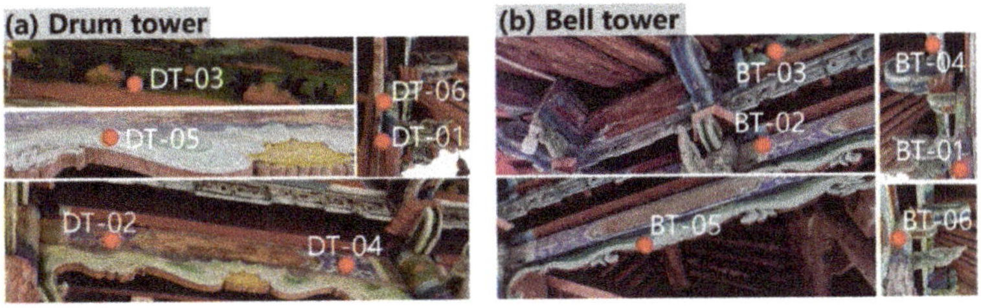

Figure 2. The Sampling position: (**a**) the Drum tower; (**b**) the Bell tower.

Table 1. Details of samples collected from the Bell and Drum Towers.

Sampling Location	**Drum Tower (DT)**					
Sampling number	DT-01	DT-02	DT-03	DT-04	DT-05	DT-06
Sampling colour	Red	White	Black	Blue	Green	Dark green
Sampling position	Beam (West)	Beam (West)	Beam (North)	Beam (West)	Queti (South)	Wooden column (West)
Sampling location	**Bell Tower (BT)**					
Sampling number	BT-01	BT-02	BT-03	BT-04	BT-05	BT-06
Sampling colour	Red	White	Black	Blue	Green	Dark green
Sampling position	Beam (West)	Beam (North)	Beam (North)	Beam (West)	Queti (East)	Wooden column (East)

Curing process was required to facilitate observations and stratigraphy analysis of the paint samples' profiles. In the first step, epoxy resin and curing solution were prepared at a ratio of 2.5:1 and poured into a 4 cm × 4 cm × 4 cm mold. After rapid stirring for 3 min until no white flocculent was visible, a portion of each sample was placed in the mold for approximately 12 h to allow the resin to fully cure. Next, a slicer was used to obtain cured sample sections with a thickness of approximately 0.5 mm, which were then sanded with sand papers and cleaned with alcohol.

To obtain accurate analysis of the type of binding medium used in the colored paintings, it was necessary to extract the binding material out from the samples. Hence, a portion of red samples, the most heavily flaked and with a larger sample size, was placed in a test tube, before adding 20% petroleum ether. Extraction involved placing the treated samples in an oscillator at room temperature for 1 h. The mixture was then separated by centrifugation and the supernatant was aspirated and placed in a test tube [20].

It was noted in the Qing building construction manual [18] that starch was one of the ingredients used in the preparation of the ground layer. Hence, in order to determine the presence of starch in the ground layers, only samples DT-02 and BT-02 (typically contained higher concentration of the white pigments) were selected for the iodine staining test. Each sample was extracted after grinding with a calcium chloride solution (Mass Fraction: 80%), where the calcium ion complexed with the hydroxyl group of the starch and allowed the starch to dissolve fully. The samples were placed on slides, before placing two to three drops of iodine into the samples. After leaving the iodine-stained samples for 5 min, the samples were observed under a microscope to detect discoloration [21].

2.2. Experimental Equipment

2.2.1. Optical Microscope and SEM-EDS

The SGO-PH80 optical microscope (ShenShi GuangGu Optical Instrument Ltd., Shenzhen, China) was used to observe the cross-sections of the solidified samples, equipped with an SGO-KK204 camera (ShenShi GuangGu Optical Instrument Ltd., Shenzhen, China). The product model of SEM-EDS was a Carl Zeiss EVO 10 (Jena, Germany) equipped with an Oxford X-act EDS (Oxford Instruments Ltd., Oxford, UK) that detected chemical elements from 4Be to 94Pu. The working distance was set to 9.88 mm and an acceleration voltage of 15 kV was used. Conductive sheet connection was applied and the magnification range was adjusted between 7 and 800 K. The sample was prepared as a solid block for SEM-EDS, with gold sprayed on the surface before analysis.

2.2.2. Micro-Raman Microscopy and FTIR Spectroscopy

Micro-Raman spectra was acquired using Thermo Fisher DXR2xi Raman imaging microscope (Waltham, MA, USA). The colored region of the sample surface was observed using a magnification of 50× objective. Laser was excited at a wavelength 532 nm. With exposure time and scanning times set to 2 s and 30 times, respectively, the recorded wave number range was adjusted from 50 to 3500 cm^{-1}.

As the Micro-Raman analysis results obtained for the red samples showed signs of fluorescence hindrance, FTIR spectroscopy was used to further support the Raman analyses. The infrared spectrometer used in this test was a Thermo Fisher Nicolet iS50 (Waltham, MA, USA) equipped with an attenuated total reflection accessory for detection within the spectral range from 650 to 4000 cm^{-1}. The attenuated total reflection (ATR) crystal type is monocrystalline diamond and the background was scanned 32 times before detection. OMSNIC software (version 8.2) was used for substance querying and baseline adjustment of the assay results. The Micro-Raman and FTIR spectra were analyzed and processed with Origin 2019 software.

3. Results and Discussion

3.1. Painting Profile Structures Analysis

A better understanding on the stratigraphy and sequential application of the Bell and Drum Towers' architectural paintings was established from the microscopic section images of the painted samples (Figure 3). The profile structures of the red samples (DT-01 and BT-01) consisted of two layers, an outer layer of red pigment and an inner layer of red and white crystalline particles interspersed within the ground layer. The white samples (DT-02 and BT-02), on the other hand, were found to have three different layers, namely an outermost white pigment layer, a blue pigment layer in the middle and a mixture of red and white crystals particles forming the ground layer. As the thickness of the white and blue pigment layers turned out to be comparable (approximately 20 μm each), the blue layer can be viewed as a form of underpainting layer for the surface white layer. The underpainting layer was mainly used to provide a transparent, coordinated base for the pigment layer. According to the painting content, the color of underpainting layer was commonly red, gray or blue, and the samples of DT-02 and BT-02 was blue mainly due to dominant color of the area in blue. In addition, the ground layer, consisting a mixture of red and white crystals particles, was mainly used as a substrate for the underpainting layer. Similar to the function of canvas, the ground layer can also indirectly serve as a protective coating for the building components. Likewise, in the case of the white samples, black (DT-03 and BT-03), blue (DT-04 and BT-04) and green samples (DT-05 and BT-05) also exhibited three layers—an outmost pigment layer, a red underpainting layer (mainly due to the dominant red color) and a ground layer of scattered red and white crystal particles.

The layering structure of the Drum Tower's dark green sample (DT-06) was similar to that of the green, blue, and black samples. However, the dark green sample of the Bell Tower (BT-06) was distinctly different from the Drum Tower's, because it contained six layers in total, where the outermost layer was the dark green pigment layer, followed by a second red underpainting layer and the third ground layer of scattered red and white crystal particles that were similar to the above samples. From the fourth to the innermost layers, the profile structure was identical with the upper three layers, namely a light green pigment layer, a yellow underpainting layer and a ground layer of white crystal particles. Thus, the above results suggested that the dark green portions of the Bell Tower might have been repainted at least once at a later period.

In general, the above analysis demonstrated the presence of red and white crystalline particles in the ground layer of all samples, with the exception of the dark green sample. By cross-referencing with historical records and the ancient Chinese building construction manuals, the features identified in this study are relatively similar to the architectural painting methodology of the Qing dynasty, where the red pigment was first applied as the underpainting layer for the columns, beams, and the brace structure (Queti), followed by the application of other pigments to the structural members of the building.

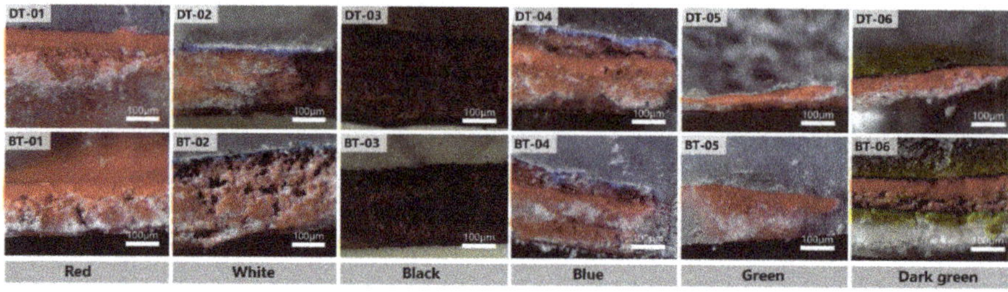

Figure 3. Microscopic section images of the painted samples.

3.2. Materials Used in Architectural Paintings

3.2.1. Pigments

Red Pigment

Iron oxide red (Fe_2O_3), mainly derived from the processing of hematite [22], was found to present in the two red samples (DT-01 and BT-01). The results obtained by EDS analysis showed that the two red samples mainly comprised of C, O, S, Ca, and Fe, with trace amounts of Na, Al, and Si (Figure 4a), thereby suggesting that the red color could be due to the presence of iron oxides. Additionally, the infrared absorption spectra obtained for the two samples contained characteristic peaks at 671 and 1109 cm^{-1} (Figure 4b). The above results corresponding to the characteristic peak of iron oxide at 671 cm^{-1} that was attributed to the Fe–O stretching vibration [23]. These analyses above confirmed that the pigment in the two red samples was iron oxide red, similar to the studies of pigments in the Dunhuang frescoes, painted sculptures in the Han dynasty tombs and architectural paintings of the Foguang Temple [4]. In addition, the two samples' characteristic peaks obtained at 711, 870, 1391, and 1793 cm^{-1} were in good agreement with the calcium carbonate characteristic peaks in the infrared spectrum, where the peak at 711 cm^{-1} was attributed to the C–O in-plane bending vibration; the peak at 870 cm^{-1} was due to the CO_3^{2-} out-of-plane bending vibration; the peak at 1793 cm^{-1} was the result of C–O symmetric stretching vibration, and the peak at 1391 cm^{-1} was caused by the C–O antisymmetric stretching vibration [24]. The presence of a trace amount of calcium carbonate could be due to previous application during repainting for color adjustment purpose or could be due to particle infiltration from the ground layer.

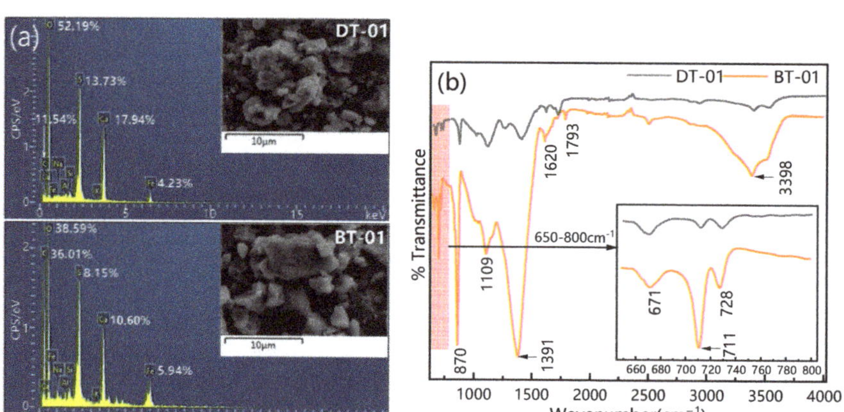

Figure 4. (**a**) Elemental analysis by SEM-EDS and (**b**) Infrared spectra by FTIR spectroscopy for the red samples (DT-01 and BT-01).

White Pigment

Chalk, comprised mainly of $CaCO_3$, also known as Baishan clay, was found to be present in the two white samples (DT-02 and BT-02). The EDS results for the white samples are shown in Figure 5a. DT-02 sample mainly composed of C, O, Ba, Ca, and S, with small amounts of Mg and Si, while C, O, Ba, Ca, S, Si, and trace amounts of Al were mainly found in BT-02. According to the elemental composition, it was estimated that Calcium or Barium salts was used in the white pigment. Further investigation into the chemical compositions of the two white samples revealed the Raman characteristic peaks at 179, 304, and 1098 cm^{-1}, and were generally in good agreement with the characteristic peaks of chalk ($CaCO_3$) [15,25] (Figure 5b). The peak at 1098 cm^{-1} was due to the C–O symmetric stretching vibration in the sample while the peak at 304 cm^{-1} arise from the stretching vibration due to the relative motion between CO_3^{2-} [24]. The above findings confirmed that the white pigment was chalk and that the presence of elements such as Al, Si, Ba, Mg, and S for both samples could have been due to contaminants adhering to the surface.

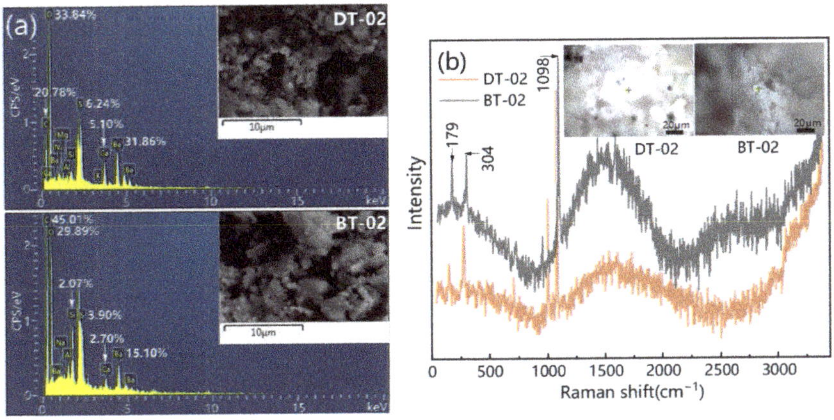

Figure 5. (**a**) Elemental analysis by SEM-EDS and (**b**) Raman spectra for the white samples (DT-02 and BT-02).

Black Pigment

Carbon black was found to be present in the two black samples (DT-03 and BT-03), generally produced from the burning of plants or animals [26]. Figure 6a presents the main constituent elements of the two black pigments. DT-03 sample consisted of C, O, Ca and small amounts of Mg, Al, Si, P, and S while BT-03 sample contained C, O, Ca, Si and trace amounts of Al and Na. From the results of elemental composition, it can be assumed that carbon or compounds of aluminum and magnesium was used in the black samples. Furthermore, the Raman spectra of the two black samples contained characteristic peaks at 1343 and 1604 cm^{-1}, which matched well with the peaks for carbon black (Figure 6b), confirmed that both samples contained carbon black [27]. Further, the absence of a characteristic peak at 960 cm^{-1} in the Raman spectra suggested that carbon black was not produced by the burning of animal fat or bone [28]. The presence of elements such as Mg, Al, Si, P, and S might be due to contaminants on the surfaces of the samples, while the high O and Ca contents might be caused by the movement of particles from the ground layer and external environment contaminants.

Blue Pigment

Lapis lazuli, one of the oldest blue pigments, was found in the two blue samples (DT-04 and BT-04) [11]. The results obtained by EDS analysis showed that the two blue samples mainly comprised of S, Ca, O, and trace amounts of Al and Si (Figure 7a). From the results of the elemental composition, the blue pigments of DT-04 and BT-04 were estimated to be Calcium or Aluminum sulfate or Silicate. Additionally, the Raman characteristic peaks of the two samples at 256, 549, 803, and 1097 cm^{-1}, corresponded with the characteristic peaks of lapis lazuli [29] (Figure 7b). These peaks of the samples were due to the S^{3-} deformation vibration, the Si–O–Si bending vibration, the Al–O symmetric stretching vibration, and the SO_4^{2-} antisymmetric stretching vibration [30]. The above results confirmed that both blue samples comprised of lapis lazuli, the same pigment commonly found in the architectural paintings of the Ming dynasty Xi'an drum tower and the Qing dynasty Linxi Pavilion in the Cining Palace [2,3]. However, by cross-referencing with the elemental results of a lapis lazuli sample from the Longxing Temple Buddha statue in Qingzhou, Shandong province [31], a higher count of O and Ca contents were noted in our samples. This could be due to the migration of particles from the ground layer or from external contaminants.

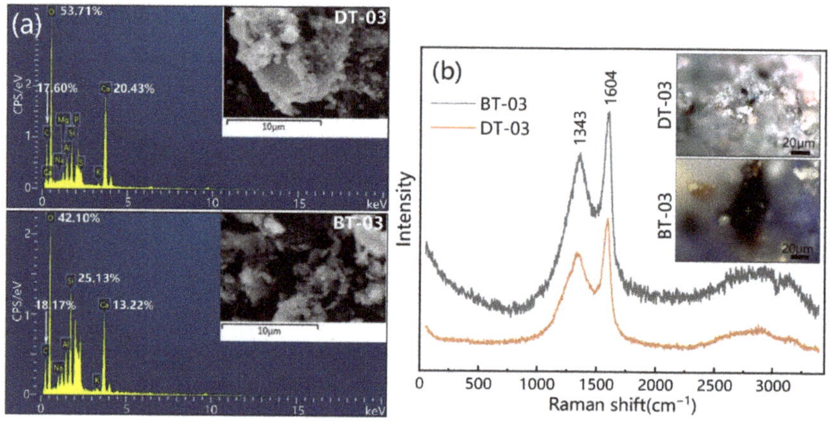

Figure 6. (**a**) Elemental analysis by SEM-EDS and (**b**) Raman spectra for the black samples (DT-03 and BT-03).

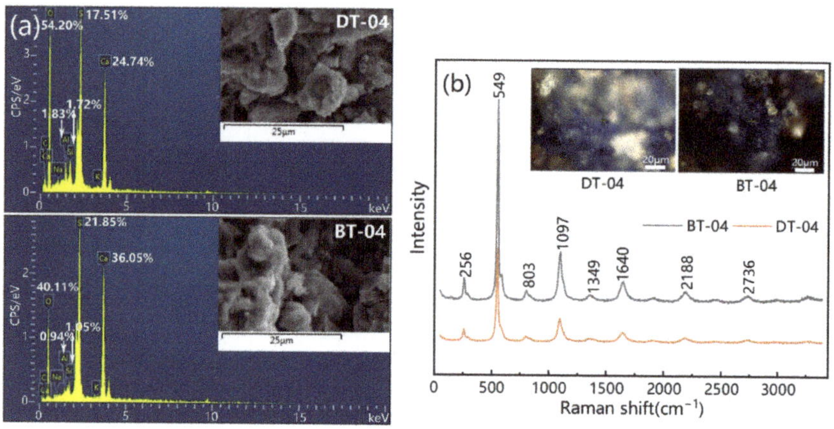

Figure 7. (**a**) Elemental analysis by SEM-EDS and (**b**) Raman spectra for the blue samples (DT-04 and BT-04).

Green Pigment

Green earth, derived from iron silicate-rich sedimentary rocks [32], was found to be present in the two green samples (DT-05 and BT-05). The EDS results for the green samples are shown in Figure 8a. DT-05 sample consisted mainly of Ca, C, O, Fe, and trace amounts of Si, Na, and Al. In the case of BT-05, high levels of Mg in addition to C, O, Ca, and Fe were observed, while no significant amount of Si, Na, and Al were detected, probably due to their low contents. According to the elemental composition's results, it was estimated that Iron salts were used in the green pigment. Furthermore, the characteristic peaks of the green samples at 142, 393, 516, 638, and 838 cm^{-1}, were in line with the characteristic peaks at 145, 399, 510, 636, and 820 cm^{-1} for green earth [33] (Figure 8b). The peak at 142 cm^{-1} was due to Fe–O stretching vibrations while the peak at 393 cm^{-1} was the result of Al–O stretching vibrations; the peak at 516 cm^{-1} was caused by the Si–O–Fe stretching vibrations whilst the peaks at 638 cm^{-1} and at 838 cm^{-1} were due to Si–O stretching vibrations and Si–O–Al stretching vibrations, respectively. Thereby confirming that the green pigment used in the architectural paintings on the Bell and Drum Towers was green earth. By comparing with the green pigments in the mural paintings on the Eastern Han tomb at Haotan in Dingbian county of Shaanxi province, the absence of potassium in our samples was presumably due to the presence of sodium instead of potassium [34].

To-date, limited literature was found on the use of green earth in ancient Chinese architectural paintings. Hence, we still do not have a clear standing why green earth was seldom mentioned in the studies relating to ancient Chinese architectural paintings. One possible speculation could be that malachite and atacamite were found to be more brightly-colored than the green earth pigment, thus making these alternatives possibly more favorable from the application and aesthetic point of view.

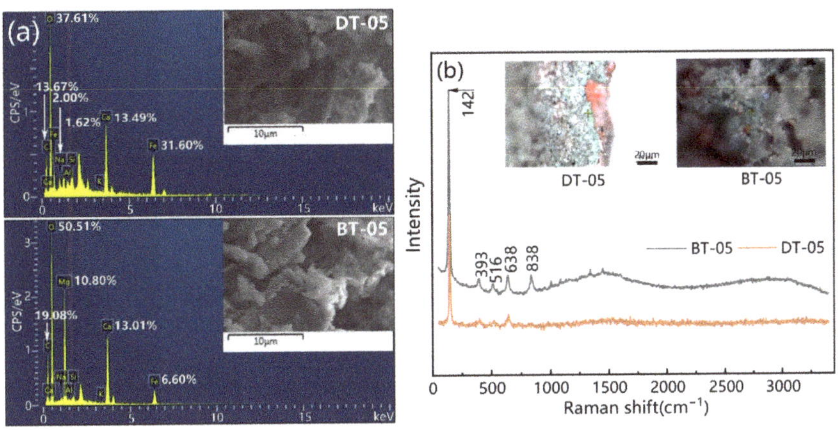

Figure 8. (**a**) Elemental analysis by SEM-EDS and (**b**) Raman spectra for the green samples (DT-05 and BT-05).

Dark Green Pigment

Figure 9a presents the main constituent elements of the two dark green samples (DT-06 and BT-06). DT-06 sample consisted of C, O, Mg, Ca, and trace amounts of Fe and Si. BT-06 contained similar constituent elements to DT-06, but with lesser amounts of Fe and Mg. Furthermore, the Raman spectra for the two samples contained characteristic peaks at 144, 838, and 1008 cm^{-1}, which matched well with the characteristic peaks of green earth in the Raman spectrum. Moreover, the characteristic peaks at 1340 and 1536 cm^{-1} were in good agreement with those of carbon black, thus confirming the presence of both green earth and carbon black pigments in the dark green samples (Figure 9b). Carbon black was probably used to tone down the brightness and color of the green pigment. The mixing of multiple pigments to adjust the color tone and brightness was also found in other studies of architectural paintings, for instance, the blue pigment in the architectural painting of Linxi Pavilion in Cining Palace was found to be a mixture of azurite, lapis lazuli, and smalt; whilst the red pigment in the architectural painting on the Foguang Temple was created by the mixing of lead red and cinnabar [3,4].

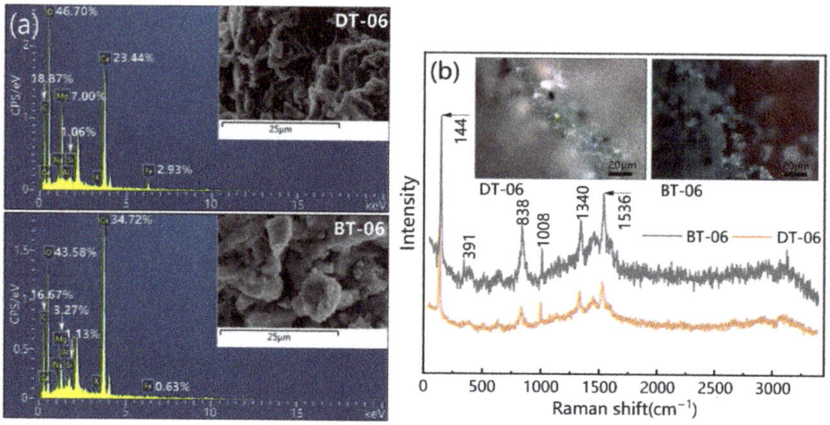

Figure 9. (**a**) Elemental analysis by SEM-EDS and (**b**) Raman spectra for the dark green samples (DT-06 and BT-06).

3.2.2. Binding Media

Results of the FTIR spectroscopy of binders in two red samples were compared with the standard spectra of binding media for animal glue, Tung oil, and egg glue (Figure 10). IR spectroscopy characteristic peaks obtained at 1456, 1075, 2854, and 2918 cm^{-1} for the two binders matched well with the three binding media's standard spectra. The peak at 1456 cm^{-1} was the result of deformation vibration for CH_2/CH_3 in lipids; the peak at 1075 cm^{-1} was due to the swing deformation vibration for CH_3; and the peaks at 2854 and 2918 cm^{-1} were the result of symmetric and antisymmetric stretching vibration for CH_2, respectively. However, further examination revealed that the two binders contained characteristic features similar to that of Tung oil where significant vibrational peaks were observed at 725, 1032, and 2958, which are assigned to the CH_2 in-plane bending vibration, the C–O stretching vibration, and the CH_3 antisymmetric stretching vibration, respectively. Animal glue and egg glue, on the contrary, will not produce significant vibrational peaks at these wave numbers. Thus, the above results confirmed that Tung oil was used as the binding medium in the architectural paintings on the Bell and Drum Towers.

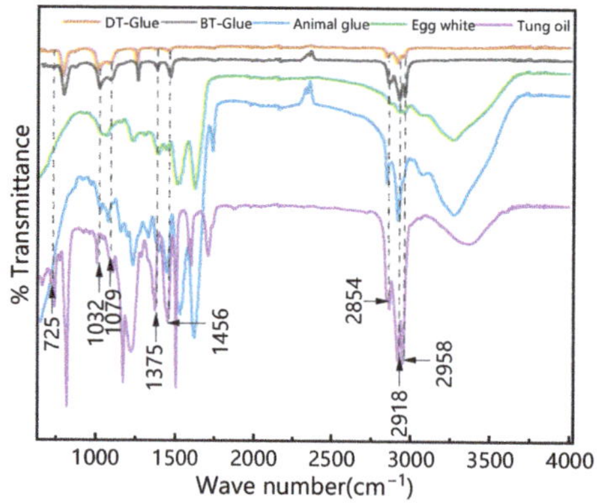

Figure 10. Infrared spectra obtained for binding media samples compared with those for animal glue, egg glue, and Tung oil.

Tung oil was a plant-based binding medium that was commonly used since ancient times. Its main chemical component fatty acid was obtained by pressing the seed from the nut of the Tung tree. This production process was recorded in ancient literature such as "*Qimin yaoshu*" and "*Jilei bian*". During the Song dynasty, the Tung oil refinement process became more sophisticated and its usage was then at its peak. In addition, Tung oil dries quickly and has good resistance to high temperatures and corrosion, as well as water. These properties are due to the presence of triple conjugated double bonds in the main component of Tung oil comprising of the tungstic acid molecule, which allows the oil to readily oxidize and polymerize in the air to form a flexible and solid Tung oil film that blocks moisture movement, resists bio-deterioration by fungi and other micro-organisms, and prevents mold and mildew. Therefore, Tung oil was widely used in architectural paintings, such as in the case of the Ming dynasty drum tower in Xi'an, Shaanxi Province [2].

3.2.3. Ground Layer

As the ground layers of the white, blue, and dark green samples (DT-02, DT-04, DT-06, BT-02, BT-04, and BT-06) are more intact, hence they were selected for EDS and Raman spectroscopy analysis. Elemental results from the EDS analysis were summarized in Table 2. The ground layers of the above six samples composed mainly of C, O, and Ca, with the exception of the dark green samples (DT-06 and BT-06) containing a small amount of Mg. The Raman characteristic peaks obtained for the six samples were similar with the chalk reference curve (Figure 11), where the characteristic peaks at 179, 304, and 1098 cm^{-1} were generally in agreement with the characteristic peaks of chalk ($CaCO_3$) [25]. With the presence of the C–O symmetric stretching vibration at 1098 cm^{-1} and the relative motion vibration between CO_3^{2-} at 304 cm^{-1} [24], the above findings further confirmed that the ground layer was chalk, and that further investigation of the dark green samples was required due to the presence of small amounts of Mg. Although chalk was found from the above results, which seemingly suggested the use of the Qing building construction manual's traditional technique for the ground layer, subsequent iodine staining test results showed no discoloration was observed in the samples (Figure 12). Additionally, the IR spectra of the samples were similar to those of the calcium chloride extract, where no C–O–H bending vibration, C–O and C–H stretching vibration were found for starch at 1090 cm^{-1}, 1130 cm^{-1}, and 2850 to 2920 cm^{-1}, respectively (Figure 13). Thus, it was confirmed that starch was not found in the ground layers of the samples and the techniques used for the preparation of the ground layer in this study was different from the method used in the Qing building construction manual.

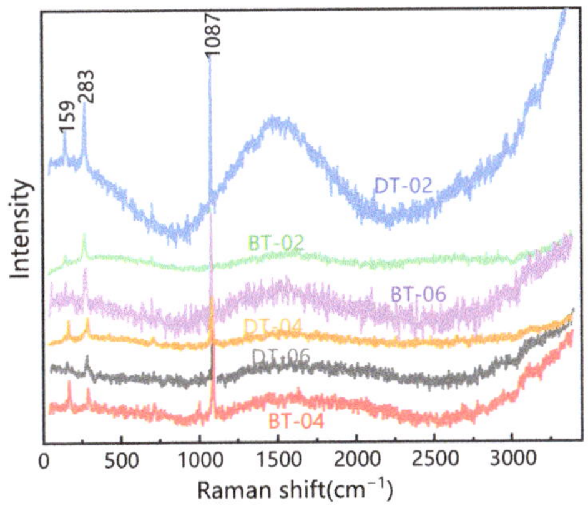

Figure 11. Raman spectra obtained for ground layer materials.

Table 2. Elemental compositions of the ground layer samples.

(wt.%) Sample \ Element	C	O	Ca	Si	Mg	Na	Al	K
DT-02	14.04	47.96	38.01	0.00	0.00	0.00	0.00	0.00
DT-04	20.53	42.74	34.51	0.00	0.97	1.24	0.00	0.00
DT-06	14.43	47.42	25.10	0.00	13.06	0.00	0.00	0.00
BT-02	16.76	47.47	35.04	0.00	0.74	0.00	0.00	0.00
BT-04	12.83	44.72	42.46	0.00	0.00	0.00	0.00	0.00
BT-06	9.94	31.50	46.42	0.00	10.32	1.82	0.00	0.00

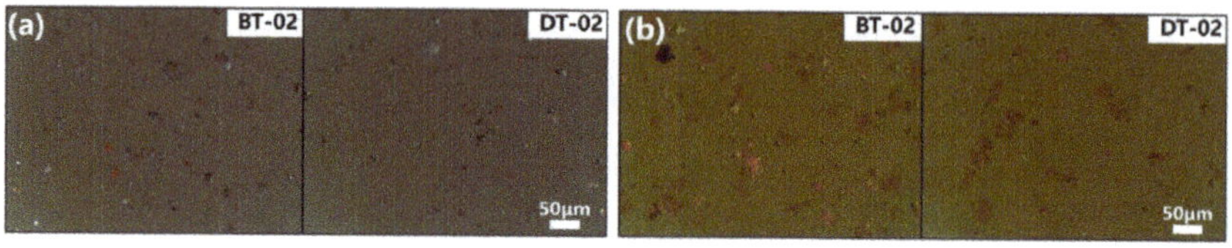

Figure 12. Micro-images obtained for the extracts of ground layer: (**a**) before iodine test; (**b**) after iodine test.

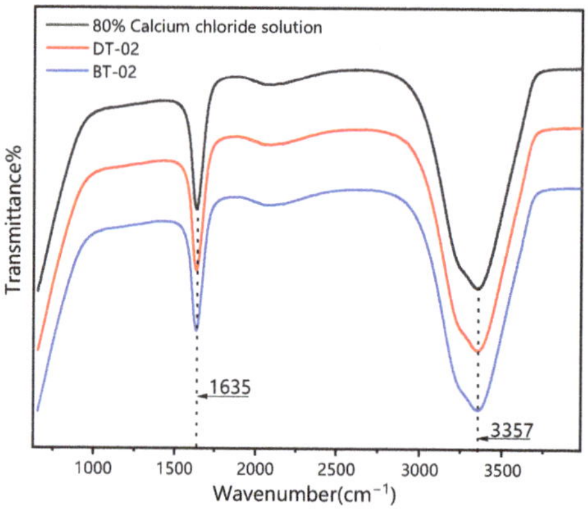

Figure 13. FTIR spectra obtained for DT-02 and BT-02.

4. Conclusions

In this study, painting profile structures, pigments, ground layers, and binding media of ancient Chinese folk architectural paintings on the Bell and Drum tower were investigated by multiple analytical methods.

Firstly, the results by optical microscopy revealed that all the architectural paintings profile comprised of pigment layers, underpainting layers and ground layers. The ground layer was found to be mainly composed of a mixture of red and white crystals particles, with the exception of the dark green sample from Bell tower that contained signs of repainting at least once at a later period. Additionally, the white crystal particles of ground layers were confirmed to be chalk, without the addition of starch.

Secondly, micro-Raman spectroscopy results showed that the pigments used in the architectural paintings of the Bell and Drum tower was iron oxide red, carbon black, lapis lazuli, chalk, green earth and dark green pigment, where the dark green pigment consisted of a mixture of carbon black and green earth. In addition, this study marks the first identification of green earth being used as a pigment in ancient Chinese folk architectural paintings. Finally, the FTIR spectra obtained from the binding media confirmed that the binding medium used in the architectural paintings of the Bell Tower and Drum Tower was Tung oil.

The findings in this study not only deepened our knowledge on ancient Chinese architectural paintings, but also highlighted the urgent need for more detailed research into ancient Chinese folk architectural pigments and binding media so as to provide more informed advice on future protection and repair works.

Author Contributions: Writing—original draft, W.Z.; Writing—review and editing, S.-Y.Y. All authors have read and agreed to the published version of the manuscript.

Funding: The study was supported by the Shaanxi Province Talent Support Grant (Grant No. 05070071240000000035) and Xi'an Jiaotong University Young Talent Support Plan (Grant No. 01190071211201020701).

Institutional Review Board Statement: Not applicable.

Informed Consent Statement: Not applicable.

Data Availability Statement: Data sharing is not applicable to this article.

Acknowledgments: The SEM-EDS, FTIR and Raman spectroscopy tests were carried out at the Instrumental Analysis Center of Xi'an Jiaotong University, China. Special thanks go to all members of the center for making the tests possible.

Conflicts of Interest: The authors declare no conflict of interest.

References

1. Li, J. The composition of pigments of decorative paintings on ancient buildings of Qufu's temple Confucius. *China Cult. Herit. Sci. Res.* **2014**, *4*, 86–89. [CrossRef]
2. Mazzeo, R.; Cam, D.; Chiavari, G.; Fabbri, D.; He, L. Analytical study of traditional decorative materials and techniques used in Ming dynasty wooden architecture. the case of the drum tower in Xi'an, P.R. of China. *J. Cult. Herit.* **2004**, *5*, 273–283. [CrossRef]
3. Li, Y.; Liu, M.Y. A Study of The Materials and Techniques Used in The Polychrome Ceiling Decoration of The Linxi Pavilion in The Garden of The Cining Palace. *Palace Mus. J.* **2018**, *6*, 45–63+159.
4. Zhang, Y.; Zhang, Z.J.; Rong, B.; Dang, H.Y. Structural survey and pigment analysis of the paint layer on the cultural relics from the Han Yangling. *Sci. Conserv. Archaeol.* **2013**, *25*, 88–92. [CrossRef]
5. Li, Y.; Wang, F.P.; Fu, X.Y.; Sun, Z.J.; Xu, Y.Q. Analysis of the pigments for smoked mural by confocal micro-Raman spectroscopy. *J. Raman Spectrosc.* **2017**, *48*, 1479–1486. [CrossRef]
6. Zhang, Y.; Wang, J.; Zhang, T. Analysis on mural structures and components of the tombs in Liao dynasty (A.D. 907-A.D. 1125). *Spectrosc. Lett.* **2015**, *48*, 732–740. [CrossRef]
7. Cheilakou, E.; Troullinos, M.; Koui, M. Identification of pigments on byzantine wall paintings from Crete (14th century AD) using non-invasive fiber optics diffuse reflectance spectroscopy. *J. Archaeol. Sci.* **2014**, *41*, 541–555. [CrossRef]
8. Iordanidis, A.; Garcia-Guinea, J.; Strati, A.; Gkimourtzina, A.; Papoulidou, A. Byzantine wall paintings from Kastoria, northern Greece: Spectroscopic study of pigments and efflorescing salts. *Spectrochim. Acta Part A Mol. Biomol. Spectrosc.* **2011**, *78*, 874–887. [CrossRef]
9. Souto, J.; Prieto, A.C.; Gutiérrez-Vicente, V. Raman analysis of gothic wall paintings in the apse of the Santiago Apóstol church in Alcazarén. *J. Cult. Herit.* **2016**, *22*, 1061–1065. [CrossRef]
10. Vandenabeele, P.; Bodé, S.; Alonso, A.; Moens, L. Raman spectroscopic analysis of the Maya wall paintings in Ek'Balam, Mexico. *Spectrochim. Acta Part A Mol. Biomol. Spectrosc.* **2005**, *61*, 2349–2356. [CrossRef]
11. Proietti, N.; Tullio, V.D.; Presciutti, F.; Gentile, G.; Brunetti, B.G.; Capitani, D. A multi-analytical study of ancient Nubian detached mural paintings. *Microchem. J.* **2016**, *124*, 719–725. [CrossRef]
12. Moretti, P.; Gallegos, D.; Marte, F.; Brunetti, B.; Sgamellotti, A.; Miliani, C. Materials and techniques of twentieth century Argentinean murals. *Procedia Chem.* **2013**, *8*, 221–230. [CrossRef]
13. Dallongeville, S.; Garnier, N.; Rolando, C.; Tokarski, C. Proteins in Art, Archaeology, and Paleontology: From Detection to Identification. *Chem. Rev.* **2015**, *116*, 2–79. [CrossRef] [PubMed]
14. Demir, S.; Şerifaki, K.; BöKe, H. Execution technique and pigment characteristics of byzantine wall paintings of Anaia church in western Anatolia. *J. Archaeol. Sci. Rep.* **2018**, *17*, 39–46. [CrossRef]
15. Želinská, J.; Kopecká, I.; Svobodová, E.; Milovská, S.; Vratislav, H. Stratigraphic EM-EDS, XRF, Raman and FT-IR analysis of multilayer paintings from the Main Altar of the St. James Church in Levoča (Slovakia). *J. Cult. Herit.* **2018**, *33*, 90–99. [CrossRef]
16. Tomasini, E.; Rodríguez, D.C.; Gómez, B.A.; Faria, D.L.A.; Landa, C.R.; Siracusano, G.; Maier, M.S. A multi-analytical investigation of the materials and painting technique of a wall painting from the church of Copacabana de Andamarca (Bolivia). *Microchem. J.* **2016**, *128*, 172–180. [CrossRef]
17. Li, J. *Ying Zao Fa Shi*; Commercial Press: Shanghai, China, 1954; Volumes 1–4. (In Chinese)
18. Liang, S.C. *Qing-Dynasty Municipal Engineering*; Tsinghua University Press: Beijing, China, 2006. (In Chinese)

19. ICOMOS China. *Principles for the Conservation of Heritage Sites in China (Billingual)*; ICOMOS China: Beijing, China, 2015. Available online: http://openarchive.icomos.org/id/eprint/1650/ (accessed on 29 January 2022).
20. Yang, L.; Huang, J.H.; Chen, X.N.; Wang, L.Q.; Wei, Y.M. PCA-LDA analysis of binders used in the theater colour painting of Jiayuguan pass based on FTIR spectra. *Spectrosc. Spectr. Anal.* **2021**, *41*, 796–800. [CrossRef]
21. Ji, L.; Shao, Z.Y. Analytical study on the painting of ancient buildings in Xi'an-Determination of starch particles in the ground layer. *Identif. Apprec. Cult. Relics* **2010**, *1*, 30–35. [CrossRef]
22. Nurdini, N.; Ilmi, M.M.; Maryanti, E.; Kadja, G. Investigation on the crystal structures of hematite pigments at different sintering temperatures. *Key Eng. Mater.* **2021**, *6139*, 20–27. [CrossRef]
23. Akyuz, A.S.; Akyuz, T.A.; Basaran, S.B.; Kocabas, I.B.; Gulec, A.B.; Cesmeli, H.B.; Uca, B.B. FT-IR and EDXRF analysis of wall paintings of ancient Ainos Hagia Sophia Church. *J. Mol. Struct.* **2009**, *924*, 400–403. [CrossRef]
24. Jin, D.L.; Yue, L.H.; Xu, Z.D. Infrared and Raman analysis of spherical $CaCO_3$ composite. *Chin. J. Inorg. Chem.* **2004**, *20*, 715–720. [CrossRef]
25. Bell, I.M.; Clark, R.; Gibbs, P.J. Raman spectroscopic library of natural and synthetic pigments (pre-1850 AD). *Spectrochim. Acta Part A Mol. Biomol. Spectrosc.* **1997**, *53*, 2159–2179. [CrossRef]
26. Goler, S.; Yardley, J.T.; Cacciola, A.; Hagadorn, A.; Ratzan, D.; Bagnall, R. Characterizing the age of ancient Egyptian manuscripts through micro-Raman spectroscopy. *J. Raman Spectrosc.* **2016**, *47*, 1185–1193. [CrossRef]
27. Choi, S.; Lee, S.K.; Kim, N.H.; Kim, S.; Lee, Y.N. Raman spectroscopy detects amorphous carbon in an enigmatic egg from the upper cretaceous Wido Volcanics of south Korea. *Front. Earth Sci.* **2020**, *7*, 349. [CrossRef]
28. Rosina, P.; Gomes, H.; Collado, H.; Nicoli, M.; Volpe, L.; Vaccaro, C. Micro-Raman spectroscopy for the characterization of rock-art pigments from Abrigo del Guila (Badajoz-Spain). *Opt. Laser Technol.* **2018**, *102*, 274–281. [CrossRef]
29. Barone, G.; Mazzoleni, P.; Cecchini, A.; Russo, A. In situ Raman and PXRF spectroscopic study on the wall paintings of Etruscan Tarquinia tombs. *Dye. Pigment.* **2018**, *150*, 390–403. [CrossRef]
30. Wang, Y. The chromophore fading and spectroscopy analysis of lazurite in annealing treatment. *Spectrochim. Acta Part A Mol. Biomol. Spectrosc.* **2020**, *247*, 119117. [CrossRef]
31. Li, Q.Q.; Zhou, L.L.; Wei, S.Y.; Ma, Q.L. Study on the Materials of the Paintings on the Head of Buddha Statue Excavated from Longxing Temple in Qingzhou, Shandong, China. *Meseum* **2017**, *2*, 28–37.
32. Fanost, A.; Gimat, A.; Viguerie, L.; Martinetto, P.; Giot, A.C.; Clémancey, M.; Jaber, M. Revisiting the identification of commercial and historical green earth pigments. *Colloids Surf. A Physicochem. Eng. Asp.* **2020**, *584*, 124035. [CrossRef]
33. Baraldi, P.; Bracci, S.; Cristoferi, E.; Fiorentino, S.; Venturi, E. Pigment characterization of drawings and painted layers under 5th–7th centuries wall mosaics from Ravenna (Italy). *J. Cult. Herit.* **2016**, *21*, 802–808. [CrossRef]
34. Fu, Q.; Xia, Y.; Wang, W.; Yang, J.; Lv, Z.; Xi, N. Study of green ground layer in an eastern Han dynasty tomb mural painting at Haotan, Dingbian county, Shaanxi province. *Sci. Conserv. Archaeol.* **2012**, *24*, 38–43. [CrossRef]

coatings

MDPI

Article

Influence of Prevailing Wind Direction on Sapping Quantity of Rammed Earth Great Wall of the Ming Dynasty

Na Su [1,*], Bo Yang [2], Wenwu Chen [3], Linrong Xu [1] and Yongwei Li [1]

1 School of Civil Engineering, Central South University, Changsha 410075, China; lrxu@csu.edu.cn (L.X.); yongweili@cus.edu.cn (Y.L.)
2 School of Civil Engineering, Henan University of Technology, Zhengzhou 450001, China; yangbo@haut.edu.cn
3 School of Civil Engineering and Mechanics, Lanzhou University, Lanzhou 730000, China; sungp@lzu.edu.cn
* Correspondence: 184801021@csu.edu.cn

Abstract: Sapping caused by prevailing wind erosion is one of the most significant factors in the deterioration of earthen sites located in Datong County, Qinghai Province, China. Long-term effects of wind may cause surface erosion, or even serious damage to the Great Wall of Ming Dynasty. Difference of sapping quantity should be attributed to variability of the prevailing wind directions. To better understand the effects of wind direction on erosion, meteorological data in the study area for fifty-two years (from 1961 to 2013) were collected and statistically analyzed. Sapping quantity of earthen structure was measured by field investigation on the Wall along the ridge whose azimuth ranges from 95°–244° and mainly concentrated in 140°–210°. Results showing obvious difference of sapping quantity could be observed at both sides of the Wall under the prevailing wind directions (ESE, SE and SSE). Further, the Wall was divided into small segments with a length of 20 m for comparison and maximum sapping quantity could be found at the Wall whose azimuth is at an angle of 30° to the prevailing wind. The aim of this study is to provide reference for the deterioration of the Wall under long-term wind pressure, and provide a targeted conservation method for earthen structure.

Keywords: earthen structure; sapping quantity; wind erosion; prevailing wind direction; structure conservation

1. Introduction

Open-air earthen heritage sites of Great Wall of Ming Dynasty, along the Silk Road in northwestern China, are highly valuable and typical representatives of historical and cultural heritage of China [1,2]. Due to its extremely high cultural and artistic value, the Ming Great Wall in Qinghai Province was listed as the 7th key national cultural relic protection units by the State Administration of Cultural Heritage of China in 2013 [3]. Generally, the Great Wall made of rammed earth, are relatively vulnerable to surrounding environment in comparison to ancient sites built by other materials, such as sintered brick, dressed stone, painted timber. For hundreds of years, earthen structures located in this district have been suffering serious deterioration induced by combination of various factors, for instance, wind flow, rain and snow, freezing and thawing, human activities [4,5]. In this context, conservation is highly urgent to mitigate or prevent the deterioration of earthen structure.

Research on the protection of earthen buildings could be traced back to the 1960s.The International Council on Monuments and Sites (ICOMOS) was established to provide a platform for the protection of earthen ruins in 1965. Thereafter, the Getty Museum, Australian Heritage Commission, International Centre for the Study of the Preservation and Restoration of Cultural Property in Rome, Dunhuang Academy and other institutions joined the preservation efforts [6]. According to field investigation and literature review, wind erosion is a typical disease of the Great Wall in northwestern China. Numerous studies have been conducted to investigate wind erosion of earthen buildings and the results indicate that common process of wind erosion could be surface stripping first, followed by sapping, finally collapse and even vanish [7]. Previous research involving wind erosion of earthen buildings mainly focus on particle size distribution, soluble salt crystallization cycles and wind-rain erosion [8–11].

It is widely accepted that sapping quantity is related to wind speed and windward of the Wall [12]. Weak regions of earthen sites would be stripped and blown away under the continuous blowing and abrading of wind, which might be further aggravated by effect of long-term and formation of empty surface. Moreover, salinized deterioration induced by soluble salt, such as sodium chloride and sodium sulfate, is significant in formation and development of sapping in earthen sites [13]. Continuous mass loss of earthen sites result from sapping should not be neglected on a long view [14–16]. Nevertheless, few studies about effect of prevailing wind direction on sapping quantity of earthen buildings are conducted during their long-term exposure to surrounding environment. Thus, it is urgent and necessary to evaluate effect of prevailing wind direction and its angle to azimuth of the Wall on wind erosion. In this study, field investigation was carried out to evaluate present situation and to obtain sapping quantity at both sides of the Great Wall in Datong County. In combination with meteorological data, especially wind regime, statistics and analysis were conducted to evaluate effect of wind erosion, focusing on prevailing wind direction and azimuth of the Great Wall on the sapping quantity [17–21]. The ultimate purpose is to further ascertain mechanism of wind erosion and to provide basis for future conservation.

2. Study Area

The Datong County is located in the northeastern Qinghai Province, in transition zone of the Tibetan Plateau and the Loess Plateau, and to the south of Qilian Mountain. The Wall in this area, originating from Yongdeng of Gansu via Huzhu of Qinghai, was built in 1572 A. D [22–26] (Figure 1). Construction background, time, and scale of the Ming Great Wall near the Xining area are shown in records by Su (1993) [27] and Yang (2016) [28]. Therefore, the Qinghai Ming Great Wall is an important part of the whole Great Wall in China according to official and historical document. Existing Great Wall in this section consists of wall with a length of 7.6 km and trench (4.8 km), 5 castles, 13 beacon towers, which is the best-preserved Ming Dynasty Wall in Qinghai, was honored as "Badaling of Qinghai" [29]. A fortress easy to defend and difficult to attack was formed by combination of the Wall and lofty and steep mountains and was listed as a red tourist spot [30].

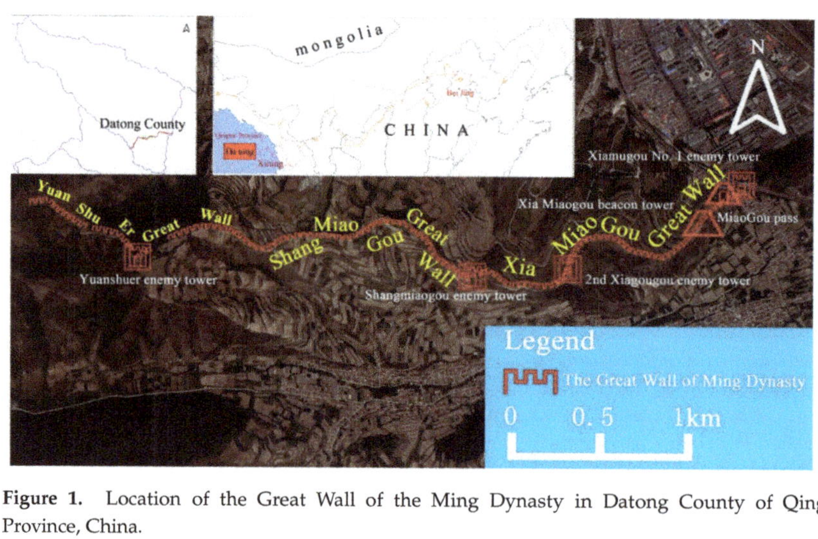

Figure 1. Location of the Great Wall of the Ming Dynasty in Datong County of Qinghai Province, China.

On account of far away from ocean, the county is characterized by continental climate. According to data from 1961 to 2013 obtained from the Qinghai Meteorological Bureau, average annual precipitation in the area is 519.9 mm, 55% of which concentrated in summer from June to August. While average annual evaporation is 1233.9 mm, about 2.4 times the precipitation. Average annual wind speed is 1.65 m/s with a maximum of 2.13 m/s in March and a minimum of 1.24 m/s in August (Figure 2a). Gale is defined in meteorology as air movement with a speed beyond 17.2 m/s. Average annual gale days is 9.3 from 1961 to 2013, with a maximum of 31 in 1965 (Figure 2b). According to monthly analysis of meteorological data from 1961 to 2013, maximum frequency of wind direction of each month is chosen as variable and shown in wind rose diagram (Figure 3). Prevailing wind direction of 52 years are ESE, SE, SSE [31,32].

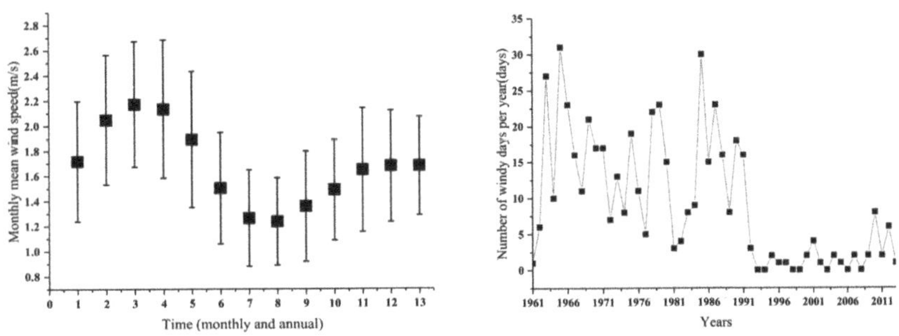

(a) (b)

Figure 2. Wind regime of study area (1961–2013). (**a**) Monthly and annual wind speed means. 1–12 indicate January to December, while 13 gives the annual average. (**b**) Number of gale windy days per year.

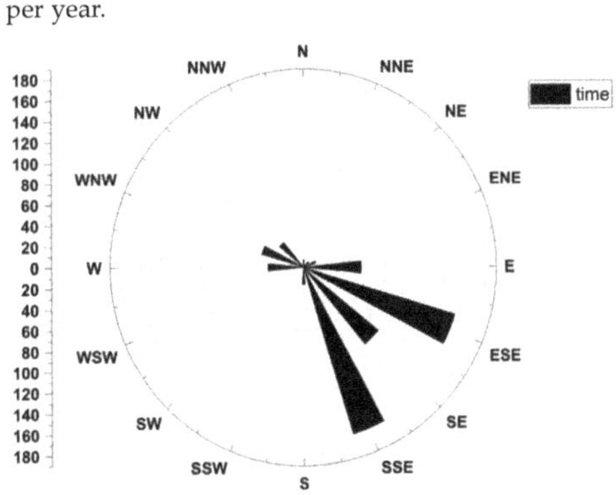

Figure 3. Wind rose diagram choosing monthly maximum frequency of wind direction (1961–2013).

As a significant part of arch-shaped defense around the Xining Garrison, Ming Great Wall is located on steep ridge of Niangniang Mountain (101°38′17.20″ E~101°41′12.90″ E, 36°55′03.50″ N~36°55′11.20″ N) with relatively slight effects of human activities and rare vegetation. Great Wall in this district, subjected to wind erosion for hundreds of years, is relatively well conserved and difficult to find in other parts of the world and of high historical and research value. Numerous studies have been conducted to investigate wind erosion of earthen sites and the results indicate that the earthen sites are vulnerable to the sapping caused by wind erosion. Therefore, it is essential to ascertain mechanism of wind erosion for effective conservation measures. As earthen heritages, the Great Wall of Ming Dynasty in Qinghai Province are essentially made of rammed earth. When wind speed exceeds threshold, the Wall body is subjected to wind erosion, mainly consisting of impact effect of sand carried by wind on surface and abrasion of eddy result from wind encountering obstacles. Impact force of wind-sand flow act directly on surface would increase with the wind speed. Further, when encountering obstacles, eddy of wind-sand flow resulted would rotate the sand particle with a high rpm, which will abrade surface of the Wall body heavily. Then, resulting surface stripping and sapping might be observed at the surface, following by the cave, groove and throughout hole developed at weak point of wall, foundation and ground. Empty face at the bottom of the Wall induced by sapping would lead to the decrease of adhesive force within rammed earth and easily stratified fall of bottom rammed earth. Besides, anti-overturning capability of the Wall would decrease with the extension of empty area, thus accelerating deterioration of the Wall.

Although built for more than 400 years ago, relevant conservation measures of the Wall in study area were not carried out until 2006 due to local economical and historical background. On the basis of massive field investigation and measurement carried out from 2008 to 2014, obvious difference in sapping quantity was found at both sides of the Wall with identical precipitation, wind field, temperature for a long time, and sapping area is mostly concentrated at upper part of the Wall [33,34]. Existing Wall body generally has a height of 0.5–5.0 m, and a width of 1.5–4.5 m at the bottom and 0.2–1.0 m at the top (Figure 4).

Figure 4. Wind erosion damage to the wall (**a**) The Wall built along the ridge; (**b**) Relative well-preserved Wall in spite of long-term nature erosion; (**c**) Wall and beacon tower on the ridge; (**d**) sapping of beacon tower; (**e**) sapping at upper part of the Wall; (**f**) measurement of sapping area; (**g**) sapping at bottom of the Wall.

3. Methods

A section of the Great Wall in Datong, under relatively well conservation state, was selected as study object to evaluate effects of wind direction on sapping quantity. Field photographs are inserted into software Computer-Aided Design (CAD in short hereafter) for the accurate measurement of sapping area for its irregular shape. A length of 20 m was chosen as reference unit in field investigation for convenience of fast measurement and calculation of sapping area in the CAD after determination of plotting scale. Twice field investigations are necessary for the accuracy and convenience of measurement of sapping. In consideration of huge change of the Wall along the ridge, length and azimuth of each segment of the Wall were measured and photographs were taken in the first investigation. In the second investigation, photographs were print and brought to field to delineate sapping area and various depth of each sapping area were measured. Then, area and mean depth of sapping could be calculated for the final sapping quantity in Excel (Figure 5).

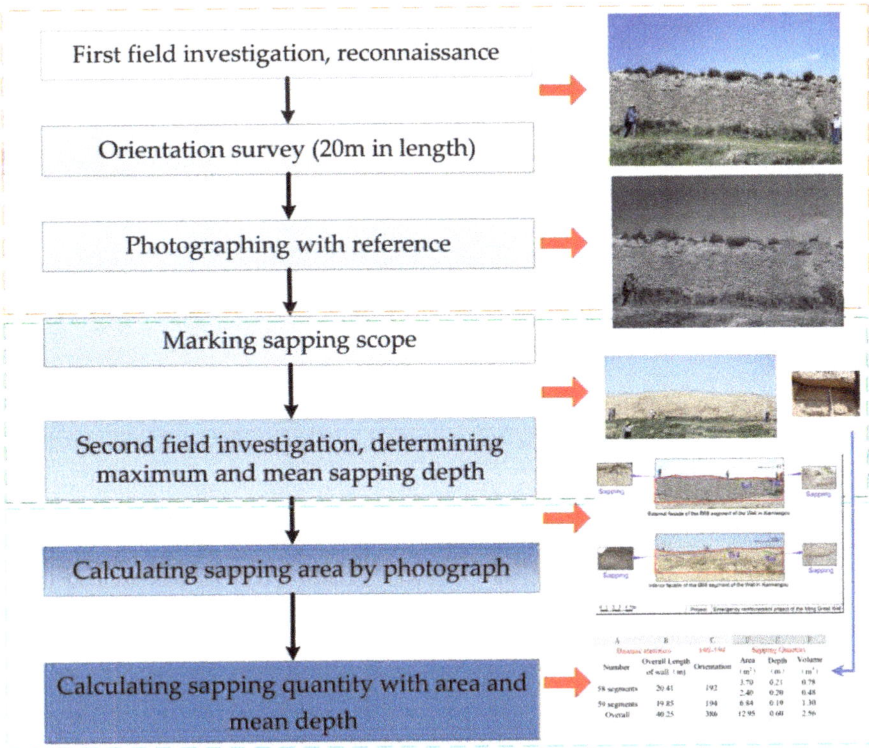

Figure 5. Calculation procedure of sapping quantity of the Great Wall.

The Wall in this area, consisting of 231 segments according to measurement method mentioned above, has a total length of 4660 m and azimuth ranges from 95° to 245°. According to azimuth of the Wall from field measurement, segments with little difference in azimuths have similar angles to prevailing wind direction. For convenience of analysis, the segments were subdivided into 31 categories by azimuth interval of 5° and numbered from 1 to 31 clockwise (Figure 6a). Length of the Wall with various azimuths concentrated from 150° to 210° were plotted in Figure 6b and detailed shown in Table 1. Sapping quantity of segments were calculated by mean sapping depth and sapping area mentioned before to improve accuracy of analysis.

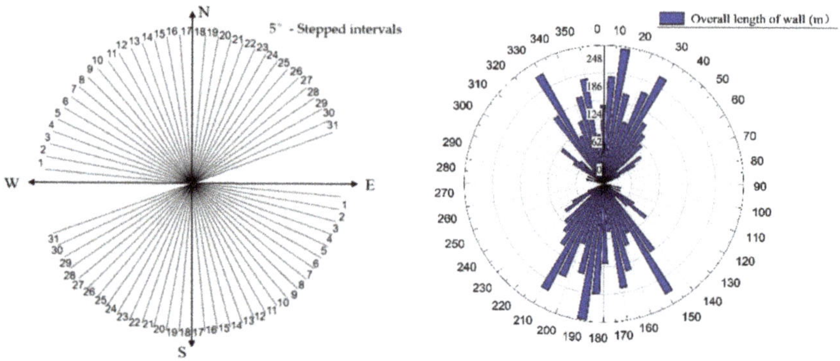

(a) (b)

Figure 6. Statistic method and results of length of the Wall. (**a**) Partition map of the north and south facades of the Great Wall. (**b**) Length of the Wall with various azimuths.

Table 1. Sapping quantity statistics of the north and south facades of the Great Wall in Qinghai.

Number	Leeward						
	Orientation (°)	Overall Length of Wall (m)	Area (m^2)	Volume (m^3)	20 m Mean-Area (m^2)	Mean-Depth (m)	20 m Mean-Volume (m^3)
1	275–279	15.89	1.46	0.33	1.83	0.23	0.42
2	280–284	40.07	1.37	0.41	0.68	0.31	0.21
3	285–289	19.95	0.00	0.00	0.00	0.00	0.00
4	290–294	39.59	0.00	0.00	0.00	0.00	0.00
5	295–299	19.93	0.00	0.00	0.00	0.00	0.00
6	300–304	80.98	1.23	0.13	0.30	0.11	0.03
7	305–309	120.21	1.78	0.23	0.30	0.13	0.04
8	310–314	71.29	0.00	0.00	0.00	0.00	0.00
9	315–319	40.14	0.40	0.10	0.20	0.25	0.05
10	320–324	182.64	1.86	0.21	0.20	0.11	0.02
11	325–329	281.14	3.53	0.70	0.25	0.20	0.05
12	330–334	117.83	0.68	0.17	0.11	0.27	0.03
13	335–339	148.27	1.20	0.18	0.16	0.13	0.02
14	340–344	201.11	5.70	0.65	0.57	0.11	0.06
15	345–349	237.95	2.76	0.32	0.23	0.13	0.03
16	350–354	110.44	0.00	0.00	0.00	0.00	0.00
17	355–359	175.75	0.00	0.00	0.00	0.00	0.00
18	0–4	240.01	1.30	0.27	0.11	0.18	0.02
19	5–9	305.06	1.36	0.24	0.09	0.22	0.02
20	10–14	205.91	5.24	0.92	0.51	0.18	0.09
21	15–19	145.07	3.88	0.66	0.53	0.17	0.09
22	20–24	222.46	10.69	1.56	0.96	0.15	0.14
23	25–29	270.38	11.76	2.57	0.87	0.22	0.19
24	30–34	162.36	3.84	0.53	0.47	0.15	0.07
25	35–39	140.03	4.43	0.91	0.63	0.21	0.13
26	40–44	119.51	0.26	0.13	0.04	0.48	0.02
27	45–49	40.11	1.56	0.26	0.78	0.17	0.13
28	50–54	101.51	20.79	1.88	4.10	0.09	0.37
29	55–59	80.56	1.12	0.14	0.28	0.14	0.04
30	60–64	40.01	0.75	0.14	0.38	0.18	0.07

Table 1. *Cont.*

Number	Leeward						
	Orientation (°)	Overall Length of Wall (m)	Area (m^2)	Volume (m^3)	20 m Mean-Area (m^2)	Mean-Depth (m)	20 m Mean-Volume (m^3)
	Windward						
1	95–99	16.21	1.44	0.23	1.78	0.16	0.29
2	100–104	40.25	12.94	2.56	6.43	0.20	1.27
3	105–109	19.34	5.23	1.10	5.41	0.21	1.14
4	110–114	40.54	9.26	1.43	4.57	0.16	0.71
5	115–119	20.36	6.49	1.30	6.37	0.20	1.27
6	120–124	80.99	18.96	4.16	4.68	0.22	1.03
7	125–129	119.82	17.54	3.94	2.93	0.23	0.66
8	130–134	71.81	10.60	2.26	2.95	0.21	0.63
9	135–139	39.34	8.61	2.03	4.38	0.24	1.03
10	140–144	184.66	60.12	14.20	6.51	0.24	1.54
11	145–149	282.94	66.09	15.57	4.67	0.24	1.10
12	150–154	122.39	37.45	9.24	6.12	0.25	1.51
13	155–159	146.93	19.00	7.72	2.59	0.41	1.05
14	160–164	202.17	46.27	15.36	4.58	0.33	1.52
15	165–169	237.42	35.88	8.43	3.02	0.24	0.71
16	170–174	108.78	7.92	2.74	1.46	0.34	0.50
17	175–179	179.70	22.09	4.55	2.46	0.21	0.51
18	180–184	247.48	32.91	4.43	2.66	0.14	0.36
19	185–189	307.79	36.28	6.20	2.36	0.17	0.40
20	190–194	206.71	25.72	5.28	2.49	0.21	0.51
21	195–199	142.64	15.69	2.88	2.20	0.18	0.40
22	200–204	226.22	34.36	4.39	3.04	0.13	0.39
23	205–209	270.25	11.39	3.45	0.84	0.31	0.26
24	210–214	164.81	37.22	6.82	4.52	0.18	0.83
25	215–219	138.72	20.44	3.80	2.95	0.19	0.55
26	220–224	122.02	8.58	1.47	1.41	0.17	0.24
27	225–229	39.57	2.38	0.39	1.20	0.17	0.20
28	230–234	105.18	1.46	0.16	0.28	0.11	0.03
29	235–239	80.81	0.70	0.11	0.17	0.18	0.03
30	240–244	39.84	0.00	0.00	0.00	0.00	0.00

4. Results and Discussion

Wind is the dominating and direct inducement of soil erosion. Degree of soil erosion induced by wind depends on speed and duration of wind [35]. A series of research have been carried out about soil erosion caused by wind and considerable consensus reached. Studies involving effect of wind on buildings could be traced back to 1960s. Thereafter, four types of methods have been developed, that is, field investigation and measurement [11], wind tunnel test, theoretical analysis and numerical simulation [26] including computational fluid dynamics [36]. Qu et al. (2007) adopted wind tunnel tests to evaluate factor of wind erosion of five earthen sites in Northwestern China [37]. Field investigation and laboratory test were employed by Wang et al. (2011) to explore mechanism of four representative diseases induced by wind erosion in Milan Ancient City [38]. On the basis of field investigation and laboratory tests, Cui et al. (2022) found that capillary process is the dominant approach of salinized deterioration effect, which would induce salt accumulation and aggravate development of sapping [39]. Although has been suffering wind erosion for centuries, the Wall in this area was relatively well-preserved due to fewer windy days and low wind speed around 1.65 m/s. Deterioration induced by low speeds wind is concentrated at lower part of the Wall, such as stripping and sapping at the bottom. Owing to higher kinetic energy and capacity of carrying sand, high speeds wind is of strong erosivity, which would lead to sapping at upper part and acceleration of deterioration of the Wall [25]. Sapping of the Wall induced by wind erosion could be attributed to stress failure after suffering wind pressure, which is closely related to speed and duration of wind. It should be noted that influence of wind direction should not be neglected, especially the area with low gale days but long duration. Generally, difference in erosion mechanism of various part of the Wall could be found according to field investigation [40].

Erosion pin method was adopted by Luo et al. (2019) in a 5 year's field monitoring to Fujian Hakka earth buildings and huge difference in sapping quantity could be found at wall with various angle to prevailing wind direction [11]. Hence, it is necessary to evaluate effect of wind direction on sapping quantity. The relation between wind direction and sapping quantity was evaluated by comparison of sapping quantity at windward and leeward and sapping area and quantity of segments with various azimuth using 20 m as reference unit length according to field investigation and data analysis. As shown in Table 1, azimuth of windward of the Wall ranges from 95° to 245° and leeward from 275° to 65°. Total sapping area of windward is 613.02 m^2 and leeward 88.95 m^2. Total sapping quantity windward is 136.20 m^3 and nearly 10 times of leeward with a sapping quantity of 13.64 m^3. Obvious difference in sapping quantity of both sides could be found although the Wall was built with identical rammed earth and located in environment with identical precipitation and evaporation. Besides, huge difference could be observed at windward of the Wall with various azimuth (Figure 7a,c). Maximum of the length of the Wall was 307.79 m with azimuth ranges from 185° to 189° (Figure 6b), while maximum of sapping quantity was 15.57 m^3 with azimuth ranges from 145° to 149°. Moreover, it could be concluded that high sapping quantity of the Wall with azimuth ranges from 140° to 154°, which is further confirmation of the relation between wind direction and sapping quantity.

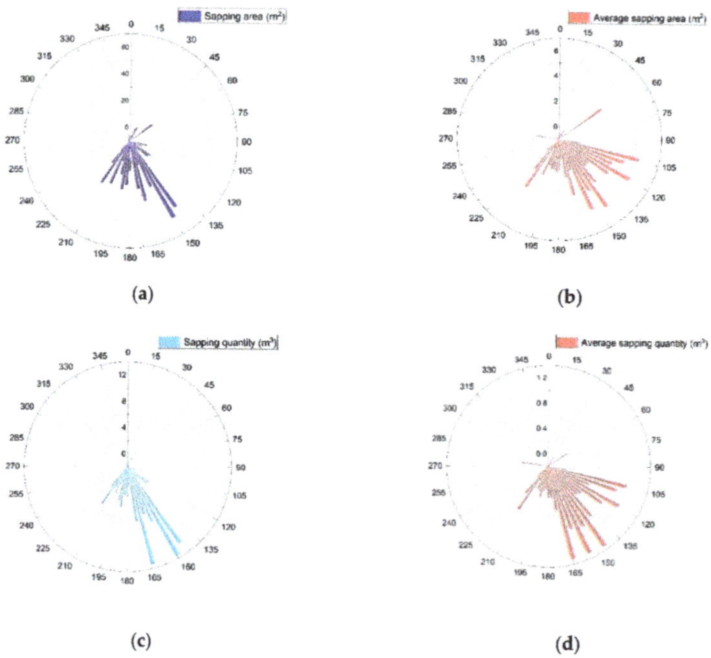

(a) (b) (c) (d)

Figure 7. Sapping of the Wall with various azimuth: (**a**) sapping area of the Wall with various azimuth; (**b**) average sapping area of the Wall with various azimuth using 20 m as reference unit; (**c**) sapping quantity of the Wall with various azimuth; (**d**) average sapping quantity of the Wall with various azimuth using 20 m as reference unit.

Richards et al. (2007) found severe deterioration in eastern facade of inner city of the Suoyang Ancient City under prevailing wind [41]. Whereas, the relationship between sapping quantity and prevailing wind direction is not analyzed further. In this paper, Figure 7b,d show average sapping area and quantity of the segment with a reference unit length of 20 m respectively, which could reveal sapping quantity of the segment with identical length. High average sapping quantity is concentrated at the segments with azimuth ranges from 105° to 165°. The highest average sapping quantity of the Wall with azimuth from 140° to 165°, is 6.72 m^3 and 33% of total and each unit segment is 1.34 m^3. The second-highest average sapping quantity of the Wall with azimuth from 100° to 124°, was 5.42 m^3 and 26% of total and each unit segment is 1.08 m^3. The third-highest average sapping quantity of the Wall with azimuth from 125° to 139°, was 2.32 m^3 and 11% of total and each unit segment is 0.77 m^3. As shown in Figure 3, prevailing wind directions in study area are, SSE, ESE, SE with frequency 167, 156, 95 respectively. Sapping quantity is of strong correlation to the prevailing wind directions. The most serious sapping could be found at windward of the Wall whose azimuth with an angle of 30° to 3 prevailing wind directions. Besides, no Wall could be observed with azimuth of 65°–94° and 245°–274° due to the long-term deterioration, which could aggravate sapping for formation of the empty area. Thus, it is necessary to focus reinforcement measures on corresponding segments of the Wall.

Fundamental objective of research about difference in sapping quantity is to optimize conservation of earthen sites and implement corresponding measures. Wind erosion of earthen sites has attracted attention in mid to late 1990s and the following is conservation measures proved effectively. Representative measures, such as surrounding environment treatment and anti-weathering consolidation, has been carried out by Dunhuang Academy since 2006 [42]. In 2012, National Cultural Heritage Administration of China issued Specifications of Investigation for Preservation Engineering of Earthen Sites (WW/T 0040-2012) [43]. Principles for the Conservation of Heritage Sites in China issued in 2015 further specify conservation measures of earthen sites, such as anchoring, grouting, back-filling, supporting [44]. This study focuses on the effect of prevailing direction on sapping quantity of the Great Wall and tries to find the area suffering severe sapping, which could improve pertinence of conservation measures.

5. Conclusions

Earthen building is a complex, open and special system in the surrounding environment and strongly influenced by it. Light, heat and water are basic elements of the environmental system and the cause of various diseases on the site. Among these, wind erosion has played a key role in deteriorating vulnerable rammed-earth sites in Northwestern China. In combination with local meteorological data, sapping quantity at both sides of the Wall were measured and analyzed mainly by field investigation. The results show that obvious difference of sapping quantity could be observed at both sides, which should be attributed to various angles of the Wall to prevailing wind direction. The following conclusions could be derived from the results of analysis:

- Sapping quantity at windward of the Wall is nearly 10 times of leeward, which is a solid proof of influence of wind direction.
- Huge difference in sapping quantity of windward could be observed for the Wall with various azimuth and highest sapping quantity could be found at windward of the Wall with an angle of 30° to prevailing wind direction.
- According to sapping level of various parts of the Wall, pertinence of traditional and mature conservation measures, would improve, which could avoid further deterioration due to inadequate conservation or waste of conservation cost because of excessive conservation.

Considering the Great Wall a significant part of world-class cultural heritage with historical, artistic and scientific values, it is of great scientific significance and application value to formulate corresponding scientific protection measures and implement reasonable and efficient protection measures [45]. There is no doubt that more multi-disciplinary work is required to predict, reduce or even prevent such deterioration of the Great Wall. This study focuses the relationship between prevailing wind direction and sapping quantity, which could provide theoretical and technical support for reinforcement of earthen heritage sites in northwestern China.

Author Contributions: Conceptualization, N.S.; Writing—original draft, B.Y.; Data curation, W.C.; Methodology, L.X. and Y.L. All authors have read and agreed to the published version of the manuscript.

Funding: The study was supported by the Henan Province Key Research and Development Program Support Grant (Grant No. 212102110200).

Institutional Review Board Statement: Not applicable.

Informed Consent Statement: Informed consent was obtained from all subjects involved in the study.

Data Availability Statement: Data sharing is not applicable to this article.

Acknowledgments: We thank the Qinghai Meteorological Bureau, China for supplying the climatic data. We would like to thank the anonymous reviewers for their valuable comments.

Conflicts of Interest: The authors declare no conflict of interest.

References

1. Li, L.; Shao, M.; Wang, S.; Li, Z. Preservation of earthen heritage sites on the Silk Road, northwest China from the impact of the environment. *Environ. Earth Sci.* **2011**, *64*, 1625–1639. [CrossRef]
2. Du, Y.M.; Chen, W.W.; Cui, K.; Zhang, K.W. Study on Damage Assessment of Earthen Sites of the Ming Great Wall in Qinghai Province Based on Fuzzy-AHP and AHP-TOPSIS. *Int. J. Archi. Herit.* **2019**, *14*, 903–916. [CrossRef]
3. Du, Y.M.; Chen, W.W.; Cui, K.; Guo, Z.Q.; Wu, G.P.; Ren, X.F. An exploration of the military defense system of the Ming Great Wall in Qinghai Province from the perspective of castle-based military settlements. *Archaeol. Anthrop. Sci.* **2021**, *13*, 1–18. [CrossRef]
4. Chen, W.W.; Zhang, Y.M.; Zhang, J.K.; Dai, P.F. Consolidation effect of composite materials on earthen sites. *Constr. Build. Mater.* **2018**, *187*, 730–737. [CrossRef]
5. Erkal, A.; D'Ayala, D.; Sequeira, L. Assessment of wind-driven rain impact, related surface erosion and surface strength reduction of historic building materials. *Build. Environ.* **2012**, *57*, 336–348. [CrossRef]
6. Chen, W.W.; Zhang, Q.Y.; Liu, H.W.; Guo, Z.Q. Feasibility of protecting earthen sites by infiltration of modified polyvinyl alcohol. Constr. *Build. Mater.* **2019**, *204*, 410–418. [CrossRef]
7. Li, Z.X.; Wang, X.D.; Sun, M.L.; Chen, W.W.; Guo, Q.L.; Zhang, H.Y. Conservation of Jiaohe ancient earthen site in China. *J. Rock. Mech. Geotech.* **2011**, *3*, 270–281. (In Chinese) [CrossRef]
8. Pu, T.; Chen, W.W.; Du, Y.M.; Li, W.J.; Su, N. Snowfall-related deterioration behavior of the Ming Great Wall in the eastern Qinghai-*Tibet Plateau. Nat. Hazards* **2016**, *84*, 1539–1550. [CrossRef]
9. Zhang, Y.; Ye, W.M.; Chen, B.; Chen, Y.G.; Ye, B. Desiccation of NaCl-contaminated soil of earthen heritages in the Site of Yar City, northwest China. *Appl. Clay Sci.* **2016**, *124*, 1–10. [CrossRef]
10. Zhang, D.X.; Wang, T.R.; Wang, X.D.; Guo, Q.L. Laboratory experimental study of infrared imaging technology detecting the conservation effect of ancient earthen sites (Jiaohe Ruins) in China. *Eng. Geol.* **2012**, *125*, 66–73. [CrossRef]
11. Luo, Y.; Yin, B.T.; Peng, X.Q.; Xu, Y.Y.; Zhang, L. Wind-rain erosion of Fujian Tulou Hakka Earth Buildings. *Sustain. Cities Soc.* **2019**, *50*, 101666. [CrossRef]
12. Su, H.; Qu, L.J.; He, K.; Zhang, Z.; Wang, J.; Chen, Z.; Gu, H. The Great Wall of China: A physical barrier to gene flow? *Heredity* **2003**, *90*, 212–219. [CrossRef] [PubMed]
13. Cui, K.; Guan, X.P.; Chen, W.W.; Chen, M.M.; Han, W.F. Effects of salinized deterioration and aeolian ullage on soils in undercutting areas of earthern ruins in arid regions (II). *J. Geotech. Eng.* **2017**, *39*, 1777–1784. (In Chinese)
14. Bui, Q.B.; Morel, J.C.; Reddy, B.V.; Ghayad, W. Durability of rammed earth walls exposed for 20 years to natural weathering. *Build. Environ.* **2009**, *44*, 912–919. [CrossRef]
15. Liu, L.Y.; Skidmore, E.; Hasi, E.; Wagner, L.; Tatarko, J. Dune sand transport as influenced by wind directions, speed and frequencies in the Ordos Plateau, China. *Geomorphology* **2005**, *67*, 283–297. [CrossRef]
16. Li, B.; Wang, L.; Liu, Y.; Liu, J. Drag distributions of non-uniform buildings from surface pressure measurements in wind tunnel. *Build. Environ.* **2018**, *143*, 618–631. [CrossRef]
17. Leenders, J.K.; Van Boxel, J.H.; Sterk, G. Wind forces and related saltation transport. *Geomorphology* **2005**, *71*, 357–372. [CrossRef]
18. Xing, F.; Mohotti, D.; Chauhan, K. Experimental and numerical study on mean pressure distributions around an isolated gable roof building with and without openings. *Build. Environ.* **2018**, *132*, 30–44. [CrossRef]

19. Du, Y.M.; Chen, W.W.; Cui, K.; Zhang, J.K.; Chen, Z.; Zhang, Q.Y. Damage assessment of earthen sites of the Ming Great Wall in Qinghai Province: A comparison between Support Vector Machine (SVM) and BP Neural Network. *J. Comput. Cult. Herit.* **2020**, *13*, 1–18. [CrossRef]
20. Meijer, M.J. A map of the great wall of China. *Imago. Mundi* **1956**, *13*, 110–115. [CrossRef]
21. Barlow, J.G. The Zhuang minority in the Ming era. *Ming. Stud.* **2013**, *1989*, 15–45. [CrossRef]
22. Liu, M.K.; Long, Y.; Zuan, X. *The Annals of Xining Wei*, 1st ed.; Qinghai People's Publishing House: Xining, China, 1993; pp. 27–43. (In Chinese)
23. Cui, Y.H. *Qinghai Economic History*, 1st ed.; Qinghai Peoples's Publishing House: Xining, China, 1998; pp. 235–247. (In Chinese)
24. Shao, M.S.; Li, L.; Wang, S.J.; Wang, E.Z.; Li, Z.X. Deterioration mechanisms of building materials of Jiaohe ruins in China. *J. Cult. Herit.* **2013**, *14*, 38–44. [CrossRef]
25. Wang, X.D.; Pei, Q.Q.; Guo, Q.L.; Li, Z.P.; Wang, Y.W.; Zhao, J.Z. Stress mechanism for the rammed layer interfaces of earthen heritage sites with different treatments. *J. Cult. Herit.* **2019**, *39*, 110–119. [CrossRef]
26. Richards, J.; Zhao, G.; Zhang, H.; Viles, H. A controlled field experiment to investigate the deterioration of earthen heritage by wind and rain. *Herit. Sci.* **2019**, *7*, 1–13. [CrossRef]
27. Su, X. *Records of Xining*; Qinghai People's Publishing House: Xining, China, 1993. (In Chinese)
28. Yang, Y. *The New Annals of Xining Fu*; Qinghai People's Publishing House: Xining, China, 2016; pp. 47–73. (In Chinese)
29. Liu, J.J.; Yan, L.; Cao, Y.C. Research on the great wall and military settlements of Xining Wei in Ming Dynasty. *Archit. J.* **2012**, *S1-07*, 30–34. (In Chinese)
30. Available online: http://m.news.cctv.com/2021/10/05/ARTIE03citWOtA4QJFwqs4b2211005.shtml (accessed on 5 October 2021).
31. Chen, W.W.; Su, N.; Yang, G. Effect of wind field on sapping quantity of earthen architecture ruins along the ridge in semi-humid area. *Chin. J. Geotech. Eng.* **2015**, *37*, 1807–1813. (In Chinese)
32. Available online: http://www.nmic.cn/data/cdcdetail/dataCode/B.0011.0001C.html (accessed on 25 March 2014).
33. Available online: http://www.gov.cn/govweb/gzdt/2007-02/13/content_525933.htm (accessed on 13 February 2007).
34. Ren, X. *Resources' Investigation Report of Ming Great Wall in Qinghai Province*; Cultural Relics Press: Beijing, China, 2012; pp. 58–149. (In Chinese)
35. Shen, Y.; Zhang, C.; Wang, X.; Zou, X.; Kang, L. Statistical characteristics of wind erosion events in the erosion area of Northern China. *Catena* **2018**, *167*, 399–410. [CrossRef]
36. Meng, F.Q.; He, B.J.; Zhu, J.; Zhao, D.X.; Darko, A.; Zhao, Z.Q. Sensitivity analysis of wind pressure coefficients on CAARC standard tall buildings in CFD simulations. *J. Build. Eng.* **2018**, *16*, 146–158. [CrossRef]
37. Qu, J.J.; Cheng, G.D.; Zhang, K.C.; Wang, J.C.; Zu, R.P.; Fang, H.Y. An experimental study of the mechanisms of freeze/thaw and wind erosion of ancient adobe buildings in northwest China. *Bull. Eng. Geol. Environ.* **2007**, *66*, 153–159. [CrossRef]
38. Wang, X.D.; Zhang, H.Y.; Yan, G.S.; Pei, Q.Q. Durability of ancient earthen architecture under wind erosion in the Milan Ancient City along the Silk Road of China. *Adv. Mater. Res.* **2011**, *163*, 3230–3236. [CrossRef]
39. Chen, G.Q.; Li, L.; Li, G.M.; Pei, X.J. Failure modes classification and failure mechanism research of ancient city wall. *Environ. Earth Sci.* **2017**, *76*, 1–15. [CrossRef]
40. Cui, K.; Zhao, X.Z.; Zhu, M.J.; Chen, W.W.; Han, W.F. Effects of salinized deterioration and aeolian ullage on soils in undercutting area of earthen ruins in arid region (III): Capillary process. *Chin. J. Geotech. Eng.* **2022**, *5*, 1–9. (In Chinese)
41. Richards, J.; Viles, H.; Guo, Q.L. The importance of wind as a driver of earthen heritage deterioration in dryland environments. *Geomorphology* **2020**, *369*, 107363. [CrossRef]
42. Cui, K.; Du, Y.M.; Zhang, Y.H.; Wu, G.P.; Yu, L. An evaluation system for the development of scaling off at earthen sites in arid areas in NW China. *Herit. Sci.* **2019**, *7*, 1–21. [CrossRef]
43. Li, Z.X.; Wang, X.D.; Chen, W.W.; Sun, M.L.; Guo, Q.L.; Zhang, J.K.; He, F.G.; Yang, S.L. *Specifications of Investigation for Preservation Engineering of Earthen Sites*; Cultural Relics Press: Beijing, China, 2012. (In Chinese)
44. ICOMOS China. *Principles for the Conservation of Heritage Sites in China (Billingual)*; ICOMOS China: Beijing, China, 2015; pp. 76–78. Available online: http://openarchive.icomos.org/id/eprint/1650/ (accessed on 29 January 2022).
45. Wang, X.D. *Philosophy and Practice of Conservation of Earthen Architecture Sites: A Case Study of the Jiaohe Ancient Site in Xinjiang*; The Peoples Press of Gansu: Lanzhou, China, 2010; pp. 1–9. (In Chinese)

coatings

MDPI

Article

Characteristics of Temperature Field of Rammed Earth Wall in Arid Environment

Qiangqiang Pei [1,2,3,4,*], Bo Zhang [1,2,3,4], Dongjuan Shang [2,3,4], Qinglin Guo [1,2,3], Jinjing Huang [1,2,3] and Jing Zhu [1,2,3]

1 Dunhuang Academy, Mogao Grottoes, Dunhuang 736200, China; zb20418@dha.ac.cn (B.Z.); gql20183@dha.ac.cn (Q.G.); hjj20384@dha.ac.cn (J.H.); zj70105@dha.ac.cn (J.Z.)
2 National Research Center for Conservation of Ancient Wall Paintings and Earthen Sites, Dunhuang 736200, China; 18368915808@163.com
3 Key Laboratory for Conservation of Ancient Wall Paintings and Earthen Sites, Dunhuang 736200, China
4 Gansu Mogao Grottoes Cultural Heritage Protection Design Consulting Co., Ltd., Dunhuang 736200, China
* Correspondence: peiqiangq@163.com

Abstract: The rammed earth walls were greatly affected by the change of temperature fields in arid environments, particularly the swell-shrink stress caused by gradient variations of the temperature fields was one of the main factors leading to surface weathering of rammed earth sites. While heat conduction, convection, and radiation were the main factors resulting in temperature variations of rammed earth walls. In this study, the thermocouple sensors were embedded in a rammed-earth test wall, to continuously monitor the temperature gradient field of the rammed earth wall from the near-surface area to the interior. The results revealed that the wall was greatly influenced by seasonal temperature fields together with significant changes in daily temperature differences. The top and the surface of the wall were affected by thermal radiation and convection, while the interior and the foundation were affected by heat conduction. The annual temperature difference reached 62.99 °C, and the maximum daily temperature difference on the surface of the wall was 24.3 °C, which mainly appeared during the spring and autumn seasons. The near-surface thickness of the temperature-sensitive area of the wall was less than 32 cm, within which the temperature varied dramatically at depths of 0–18 cm. The temperature variations at depths of 18–32 cm were somewhat significant with no obvious changes at depths exceeding 32 cm. These trends indicate that the rammed earth wall has an outstanding function of thermal preservation and insulation. This study is expected to be of guidance and reference for multi-field coupling environmental condition setting for surface weathering of rammed earth site bodies, stress and strain caused by temperature field, surface weathering mechanism and strengthening technology as well as the related researches of modern rammed earth building designs.

Keywords: rammed earth sites; temperature field; gradient law; arid environment

1. Introduction

In recent years, as human awareness of the environment has been improved and problems related to housing energy efficiency and environmental aspects have been concerned, rammed earth structures have attracted particular attention owing to their special features such as poor heat conduction, good thermal insulation, and low energy consumption [1,2]. The intense thermal energy transference in a certain range of depths from the surface of the rammed wall ensures relatively constant temperature and humidity for inside parts of buildings [3]. However, conservators of rammed earth sites have focused mostly on the damage from dramatic temperature fluctuations on the surfaces of such sites. Matthew Hall et al. believe that thermal conductivity materials increased linearly with their saturation ratio [4]. Zheng Long et al. adopt the method of implanting silicon semiconductor sensors in the wall to monitor the wall temperature and find that the temperature gradient varied clearly from the wall surface to the center [5]. Globally, cultural and heritage sites are carefully preserved on all continents, and more than one-third of these sites are earthen sites [6,7]. A large number of culturally rich earthen sites are preserved in northwest China [8]. Under the influence of long-term natural and human factors, these sites have suffered from different types of deterioration [9]. The weather in northwest China is extremely variable, with the annual maximum temperature difference reaching 80 °C; the daily maximum temperature difference approaches 30 °C in some areas. Earthen sites in some locations are covered in snow for long periods during winter and directly exposed to sunlight in summer. Therefore, the temperature fields between the superficial and internal layers of these sites vary greatly, thereby causing long-term repeated expansions and contractions that can gradually affect the structures of the superficial layers [10]. This is the main driving force for the intensification of surface weathering, flaking, stripping, and foundation undercutting damages in earthen sites (Figure 1). Jenny Richards et al. established a preliminary model for sand degradation of earthen sites in the arid area [11]. Du Yumin et al. took the Great Wall of Qinghai as the object to evaluate the damage degree of the sites by means of a neural network and fuzzy analytic hierarchy process and ranked the main influencing factors [12,13]. Although the influence of the natural environment on the destruction of earthen sites is a multi-field coupling effect, the main causes are considered to be water and temperature variations. Fodde, Enrico et al. consider that groundwater is the key driver for salt-eroded sites [14]. Sun Manli and Susana Serrano, etc. believe the rammed earth wall was easy to weather and fall off under the action of water and heat [15,16]. The changes in the temperature fields at the sites are greatly affected by the seasonal climate, which is the necessary driving force for the movement of water and salt as well as the expansion and contraction of the earthen sites; these variations also render the rammed earthen sites more prone to damages [17–19]. Recently, Shen et al. [20] studied the model of random solar radiation energy, especially the processes of storing and releasing energy, which can help reduce fluctuations in indoor temperatures and improve the level of thermal comfort. Taylor and Luther [21], Soudani et al. [22] and Fernandez et al. [23] believe that earthen materials, especially the rammed earth, are able to store large amounts of thermal energy and display latent heat owing to the liquid-to-gas phase change within the pores. However, as early as the creation of Yueling in the Book of Rites [24], there were records stating "in the month of Meng Chun, the weather falls, the earth's atmosphere rises, the heaven and the earth are in harmony, and vegetation sprouts," which accurately explains the changing law of the Earth's temperature field in spring. In addition, a great deal of importance has always been attached to agriculture in China. As early as in the Western Han Dynasty's agricultural work—the Book of "Fan Sheng Zhi Shu" [25]—the Chinese people have demonstrated a profound understanding of the relationship between seasons and soil, in order to guide agriculture production. The changing characteristics of the temperature fields and how they are affected by the seasons of spring, summer, autumn, and winter are explained in the book "Tian Ji Su Shu" [26]. All these quotes from literature reveal the existence and belief in the basic theory of relative latent heat.

(a) Rain wash

(b) Site surface snow

(c) Weathered surface

Figure 1. Influence of natural conditions.

Extant research has noted that long-term dramatic temperature changes will cause expansion and contraction of the site body [27], as well as water and salt migration, water solidification [28,29], and volume expansion caused by water vapor [30]. In the process of repeated cyclic expansions and contractions, water and salt transport are enhanced (unsynchronized expansion and contraction of the earthen body, dissolution of soluble salts, crystalline expansion, re-shrinkage, and re-expansion), which leads to an obvious reduction in surface strength at the site and the appearance of multiple surface damages such as disruption, powdering, flaking, and peeling [31]. Among these causes, the sharp gradient changes in temperature fields are considered the internal driving force, and the changes to the solid, liquid, and gaseous phases of water are the main modes of energy storage and release (Figure 1).

In the past few years, increasing numbers of researchers and conservationists have begun to study the influence of temperature fields on sites. Some studies have concluded that strong temperature changes are one of the main causes of weathering on the surfaces of the earthen sites [32–34]. In particular, sharp rises and falls in the ambient temperature can rapidly increase or decrease the spacings between soil particles. Under repeated cycles of this process, the cohesion between soil particles decreases, soil surface becomes loose, and soil particles can even be loosened under external forces [28], the change of temperature field with severe gradient plays an important role [5]. Therefore, changes in the temperature fields with sharp gradients play an important role in the structural integrity at these sites.

However, owing to the influence of the temperature field, monitoring the temperature gradients of rammed earthen sites has always been a challenging problem. To solve this, it is necessary to further characterize the progressive gradient change processes of temperature fields, from the shallow surfaces to the insides of earthen sites. This study represents efforts toward the evaluation of the gradient temperatures of the surface and internal structure of a rammed earthen wall, which are monitored by embedding 5TM soil moisture temperature sensors and adopting traditional wall and atmospheric temperature testing equipment. The test methods for the superficial layer and the internal temperature fields of large-volume rammed earth walls are also summarized in this paper. The variation characteristics and gradient laws of the 24 h wall temperature field of rammed earth sites in an arid environment were preliminarily characterized, including the gradual changes in the characteristics and evolution laws of the wall temperature fields in spring, summer, autumn, and winter. These variations were found to be extremely similar to the seasonal influence proposed in the ancient texts. This paper also proposed an empirical formula to describe the changes in the walls at rammed earthen sites with respect to atmospheric temperature. Further, the temperature field, stress field, and coupling relationships, as well as the weathering mechanisms that affect the surfaces of earthen ruins, were also investigated in this research.

2. Research Aim

The expansion and contraction stress caused by the gradient change of temperature field is one of the main factors leading to surface weathering of rammed earth sites. Therefore, the variation range and amplitude of the temperature gradient on the surface of rammed earth walls are the main indicators that lead to surface weathering. This paper discussed the influence of daily temperature difference and annual temperature difference on the walls through monitoring and revealed the sensitive areas and main seasons that affected the wall surface temperature, which provides a scientific basis for the prevention and control of surface weathering of rammed earth sites.

3. Methods

3.1. Test Wall Ramming

(1) Basic properties of test soil

The experimental soil was taken from platform soil, which is similar to the rammed soil of the Western Xia Mausoleum in Yinchuan, China, 80 km from No. 9 Western Xia Mausoleum. It belongs to loess, which is odorless and contains a small amount of flaky crushed stone. The mineral composition and particle distribution of the soil sample are given in Table 1.

Table 1. Basic properties of test soils.

Mineral Components		Distribution of Particle Size		Basic Physical Properties	
Component	Content (%)	Particle Size (mm)	Content (%)		
Quartz	30	20–10	1.71	Moisture content (%)	1.84
Calcite	25	10–5	4.24	Gravity (Gs)	2.72
Dolomite	7	5–2	5.68	Liquid limit (%)	25.9
Feldspar	9	2–1	0.27	Plastic limit (%)	17.6
Illite	21	1–0.5	2.89	Plastic index (IP)	8.3
Chlorite	7	0.5–0.25	1.77	Non-uniformity coefficient Cu	8.27
-	-	0.25–0.075	28.63	Curvature coefficient Cc	1.17
-	-	<0.075	64.66	-	-

(2) Rammed test wall

The test wall was rammed using traditional techniques [18]. The rammer had a mass of 6.9 kg with a diameter of 12 cm. The thickness of rammed pave soil is 12 cm, and each layer has been rammed six times for testing; the rammed test wall has a length of 3.4 m, the height of 2 m, and bottom width of 1.05 m, with a north to south trend; after the ramming, the wall was trimmed to a length of 3.0 m, the height of 2.3 m, and bottom width of 1.1 m (Figure 2).

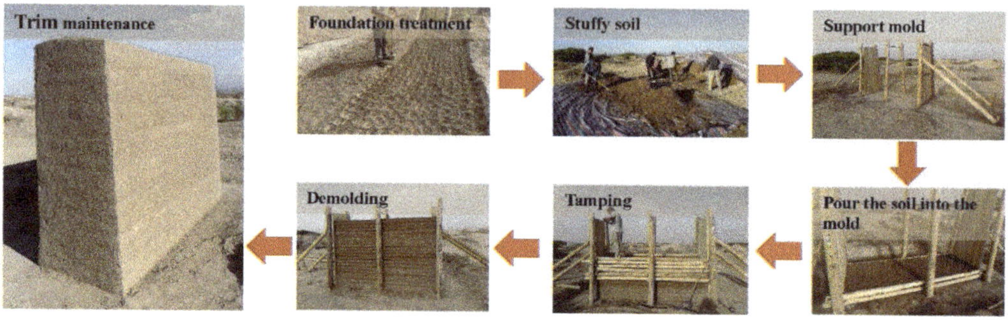

Figure 2. Testing wall tamping process.

3.2. Monitoring Equipment and Installation of the Test Wall

3.2.1. Monitoring Instruments

(1) 5TM temperature and humidity sensor

The 5TM sensor for soil temperature and humidity, produced by the Decagon Company (Lagos, Nigeria), was used for measurement. The sensor has the advantages of a small volume, strong corrosion-resistance, high measurement accuracy, and simple operation. The temperature measurement range was −40 to 60 °C, the humidity range was 0% to 100%, the measurement accuracy was 1 °C, and the measurement resolution was 0.1 °C. The temperature value of the soil was obtained by the thermistor on the sensor surface, and the temperature data measured by 5TM were collected by an EM50 collector (Figure 3).

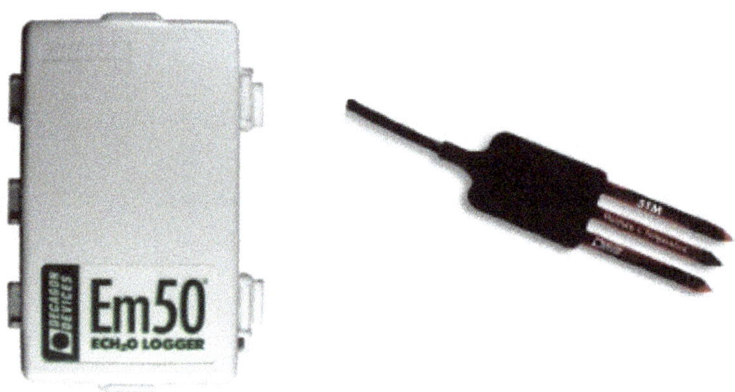

(**a**) EM50 collector (**b**) 5TM soil temperature and humidity sensor

Figure 3. Testing instrument.

(2) Weather station

A meteorological station was set up at a distance of 10 m from the test wall to collect environmental data, such as ambient temperature, atmospheric relative humidity, atmospheric pressure, wind, rain, and dew, and data were sampled every five minutes. The collected data were compared and analyzed with data measured in the wall to provide basic information for analyzing the influence of the atmospheric environment on the temperature field of the wall (Figure 4).

(**a**) Plane distribution map of the test area (**b**) Environmental data collection

Figure 4. Meteorological station.

3.2.2. Embedding Method for the Monitoring Equipment

(1) Sensor check

One end of the 5TM soil moisture temperature sensor was inserted into the collector and then connected with the computer through the matching data line. The time and date were corrected, and the parameters and numbers were then set according to the test requirements and instrument instructions.

(2) Sensor positioning

The density of the soil for the temperature field test was 1.70 g/cm^3, and the sensors were arranged inside the test wall in a total of 11 layers from bottom to top. Each layer, from A to D, was equipped with eight sensors that were symmetrically distributed around the central axis. Each layer from E to I was equipped with seven sensors that were symmetrically distributed around the central axis. The distance between the first sensor on both sides of the periphery of the site body and the surface of the test wall was 50 mm; the distance between the first sensor and the adjacent second sensor was at least 70 mm, and the rest of the sensors were all arranged at a distance of 100 mm. The sensors at the top and bottom of the test wall were arranged densely, the positions of the embedded sensors were marked, and the sensors were then placed on the marked points in sequence and recorded. The sensor arrangement is shown in Figure 5.

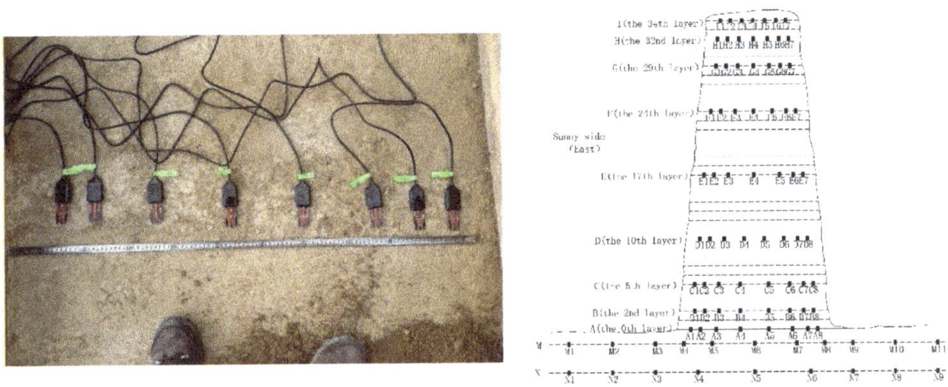

(**a**) 5TM sensor is laid out on site (**b**) 5TM sensor layout inside the wall

Figure 5. 5TM soil moisture temperature sensor layout.

(3) Sensor embedding

The virtual soil was paved and compacted under the sensor, to ensure that the sensor was in close contact with the soil. The sensor was fixed by hand, and the soil was then paved onto it to ensure that the sensor would not move when the virtual soil was paved.

(4) Ramming paved soil

The sensor leads were first fixed and then ordered. A small amount of experimental soil was then lightly pressed onto the leads to further fix them. When the sensor and leads were completely fixed, the soil was spread from the ground to the top surface of the test site body, from layer A to I. The test wall was divided into 34 layers; layer A was on the ground, there were two layers from layer A to layer B, three layers from layer B to layer C, five layers from layer C to layer D, and two layers between each layer from layer E to layer F, three layers from G to H, two layers from H to layer I, and one layer above I. The thickness of each layer was 59 mm. After the soil was paved, it was rammed six times with a rammer.

(5) Data acquisition and processing

Data were acquired every five minutes and processed to establish the position coordinate system of the test wall, divide the grid, and then draw a temperature isotherm diagram. The isotherms of the test wall at these two moments were drawn using a linear difference triangulation method in the Surfer software, in order to ensure the accuracy of the isotherm data.

4. Results

Owing to the periodic changes in the seasons of the year, the monitoring data of the temperature field of the test wall [17,19] were analyzed for one full year after the test wall was completely air-dried. In this study, the ambient temperature and test temperatures were measured at different positions on the test wall from 1 March 2017 to 1 March 2018, to comprehensively analyze how the wall was affected by the environment.

4.1. Seasonal Variation of Wall Temperature Field

Through continuous monitoring of the temperature field of the test wall, it was found that the lowest point and highest temperatures occurred at around 04:00 and 16:00 over one day, respectively. Figure 6 shows the temperature field isotherms at 04:00 and 16:00 for 12 consecutive months.

It can be seen from the isotherm at 04:00 in Figure 6 that the overall surface temperature is lower than the internal temperature of the test wall. With seasonal changes, the internal temperature was 1.5–8.5 °C higher than that of the surface. The maximum temperature difference (8.5 °C) occurred in April, whereas the minimum temperature difference (1.5 °C) occurred in August. The isotherms were distributed symmetrically along the central axis of the test wall. The ambient temperature was lower than the internal temperature of the test wall at 04:00 when there was no solar radiation. The test wall presented good uniformity in all directions and provided thermal insulation. The heat was uniformly dissipated from the test surface, and the overall surface temperature of the test wall was lower than its internal temperature.

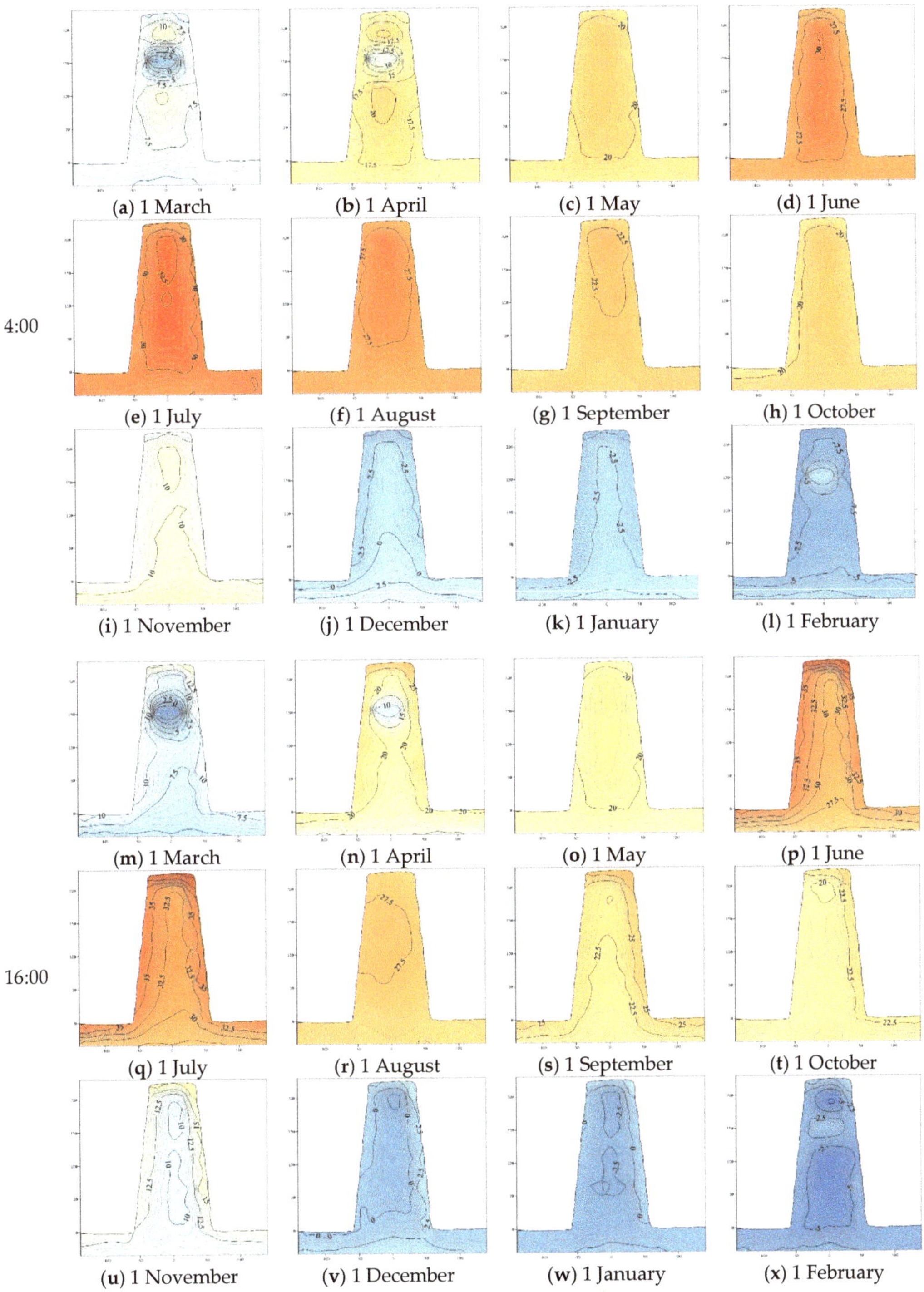

Figure 6. Trend charts showing changes in temperature field isotherms at 4:00 and 16:00 in one year.

In addition, as the foundation of the wall was affected by geothermal energy and the top was affected by solar radiation, the overall temperature of the wall changed relatively rapidly. However, the temperature in the range from 1.0 to 1.5 m from the middle-upper part of the wall was relatively constant. In January, the temperature of the entire wall was below zero, and the temperature decreased gradually from inside to outside, displaying a parabolic profile. In February, the temperatures at the top and foundation of the wall were higher than in the middle. In March, the temperatures of the wall were in the following order from high to low: top, bottom, and middle. In April, this changed to the bottom, top, and middle, and from May to September, the order changed to the middle, top, and bottom. The bottom temperature gradually decreased, and the area of the middle-temperature circle gradually shrunk. From October to December, the bottom temperature gradually rose, and the wall temperatures were in the following order from high to low: middle, bottom, and top.

The isotherm at 16:00 shown in Figure 6 reveals that the distributions of the internal and surface isotherms of the test wall are opposite to those at 4:00. With seasonal changes, the wall surface temperature was 1.5–12.5 °C higher than that inside. The maximum temperature difference was 12.5 °C (in July), and the minimum temperature difference was 1.5 °C (in August). The isotherms were asymmetrically distributed, and the top temperature was higher than that of the lower area of the test wall. With the change in the azimuth angle of solar radiation, at 16:00, the west facade of the wall was directly affected by the perpendicular incidence of the sun rays, and the temperature on the western surface increased rapidly. This asymmetric radiation caused the temperature of the west facade to be lower than that of the east facade; as the top of the test wall was exposed to direct solar radiation from the beginning to the end, its temperature was the highest. This was particularly true in July when the top temperature of the test wall could reach 46.99 °C.

4.2. *Seasonal Diurnal Variation Characteristics of Wall Temperature Field*

According to the changing trend of the temperature field isotherms in the above year, the temperature of the shallow surface layer of the test wall changed with the changes in the ambient temperature, and when the ambient temperature changed rapidly, the temperature of the shallow surface layer of the test wall also changed rapidly. The change in ambient temperature had little influence on the temperature change inside the test wall, and it decreased stepwise from the outside to the inside. The area where the wall temperature changed by more than 10 °C across the four seasons was termed the sensitive area; the area where the wall temperature changed between 5 and 10 °C was termed the transition area; the area where the wall temperature changed between 1 and 5 °C was termed as the stable area, and the area where the wall temperature changed by less than 1 °C was termed as the constant area. Considering the climate change in the four seasons, efforts were made to reduce the influence of extreme weather changes such as clouds, rainfall, wind, and snow on the solar radiation received by the test wall. To this end, monitoring data of the representative spring equinox, summer solstice, autumn equinox, and the best sunshine days near the winter solstice in the four seasons were collected for comparative analysis.

4.2.1. Spring

As seen in Figure 7, sunshine was the strongest from 00:00 on 22 March to 00:00 on 23 March 2017. Starting from 00:00 on 22 March, the data were collected every 4 h to complete the isotherm diagrams, as shown in Figure 8.

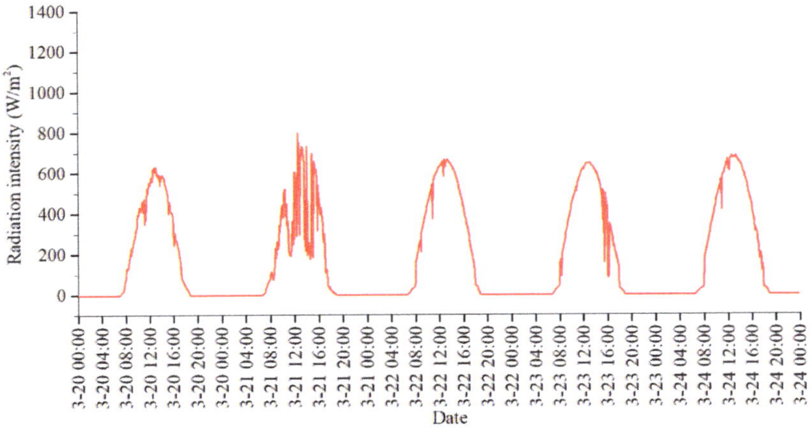

Figure 7. Selection of optimum days for spring illumination.

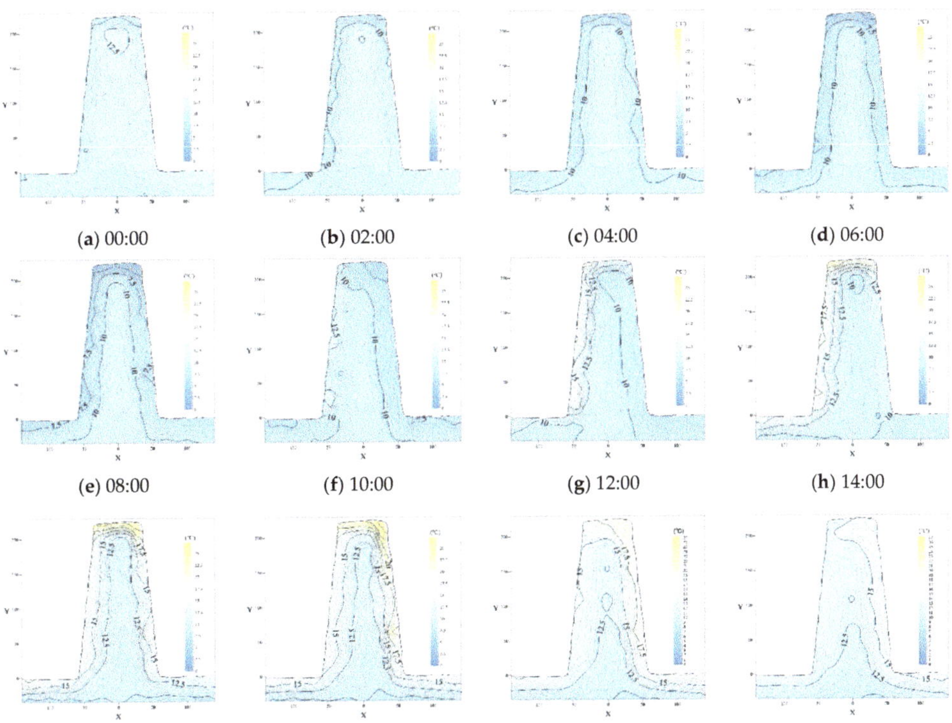

(**a**) 00:00 (**b**) 02:00 (**c**) 04:00 (**d**) 06:00

(**e**) 08:00 (**f**) 10:00 (**g**) 12:00 (**h**) 14:00

(**i**) 16:00 (**j**) 18:00 (**k**) 20:00 (**l**) 22:00

Figure 8. Temperature variations in the test wall on 22 March.

As can be seen from Figure 8, the temperature inside the test wall is almost constant for one day in spring (March), whereas the surface temperature changed greatly. The temperature field change of the wall surface can be divided into three stages: the temperature gradually fell down to the lowest from 00:00 to 08:00, rose up to the highest from 08:00 to 16:00, and dropped slowly from 16:00 to 24:00. These observations show that the wall temperature decreased from 16:00 to 08:00 on the next day. To be specific, it decreased slowly from 16:00 to 20:00, and then decreased rapidly from 20:00 to 08:00 on the next day. From 08:00 to 14:00, the temperature of the wall gradually increased, and the highest temperature in this stage occurred in the top corner area on the east side of the test wall, reaching 17.5 °C. The wall temperature was the highest between 14:00 and 16:00, and the hottest area of the test wall gradually moved from the east to the west, i.e., to the corner area of the west façade; and the maximum temperature was 26 °C.

According to the above changes in the different temperature fields, the wall was divided into four different areas: the sensitive area, transition area, stable area, and constant area, among which the sensitive area was divided into the strong sensitive area and sensitive area. The strong sensitive area is the red-filled part on the west side of the top of the wall in Figure 9, and it shows the largest temperature change. The wall temperature at 16:00 was 20 °C higher than the temperature at 08:00; this area was located 16.17 cm from the western side and 20.81 cm from the top of the wall. The sensitive area is the orange-colored part in Figure 9, which occupies the top of the wall and the surface layer on the upper half of the west side of the wall. The thickness of this top layer was 14.55 cm, its thickness on the west side was 10.20 cm, and its height was 82.24 cm. The temperature at 16:00 was 12.5 °C higher than that at 08:00. The transition area is the green-filled part in Figure 9. Compared to that at 08:00, the temperature rose by 5 to 8 °C. The thickness of this area was about 20 cm, with the thicknesses of 21.28 cm on the east side, 12.26 cm on the top, and 22.85 cm on the west side. The stable area is colored cyan (Figure 9). Compared with the temperature at 8:00, the temperature rose by 1–3 °C; the east side was 42.11 cm away from the wall surface, and the west side was 25.26 cm away from the wall surface. The constant region is the blue-filled part (Figure 9), having a roughly triangular shape. The height of the constant region was 102.96 cm, and the temperature remained almost unchanged within this area.

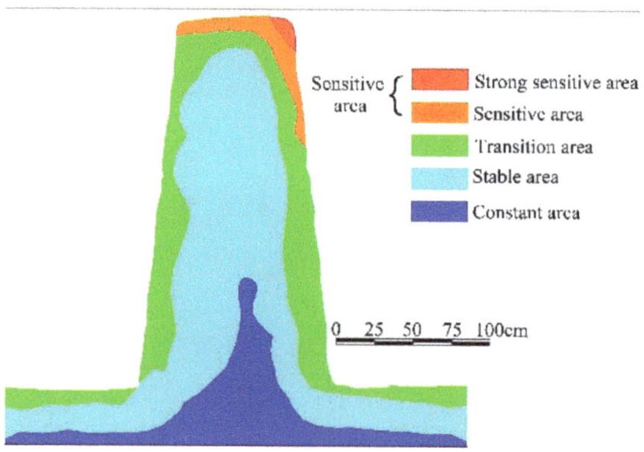

Figure 9. Temperature field of test wall in spring.

To summarize, the spring temperature field was divided into four areas, as shown in Figure 9 and the average temperature change in the sensitive area was above 10 °C. The temperature change in the transition area was 5–8 °C, and that in the stable area was 1–3 °C; the temperature in the constant area remained unchanged compared, as opposed to the other time periods (Figure 9).

4.2.2. Summer

As can be seen from Figure 10, the sunshine was brightest from 00:00 on 16 June to 00:00 on 17 June 2017. Starting from 00:00, data were collected every 4 h to complete the isotherm chart, as shown in Figure 11.

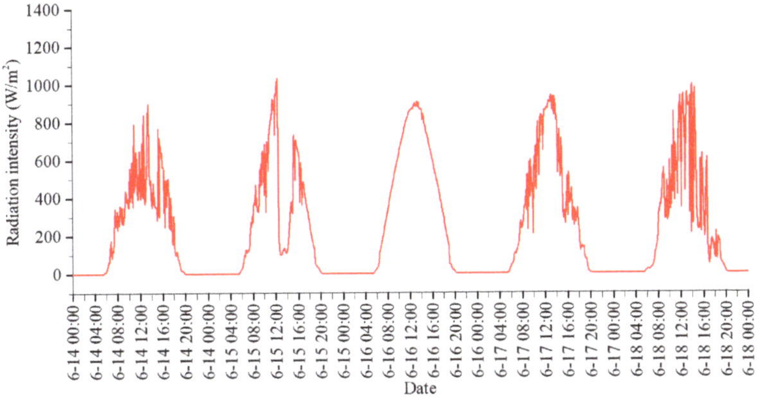

Figure 10. Selection of optimum days for summer illumination.

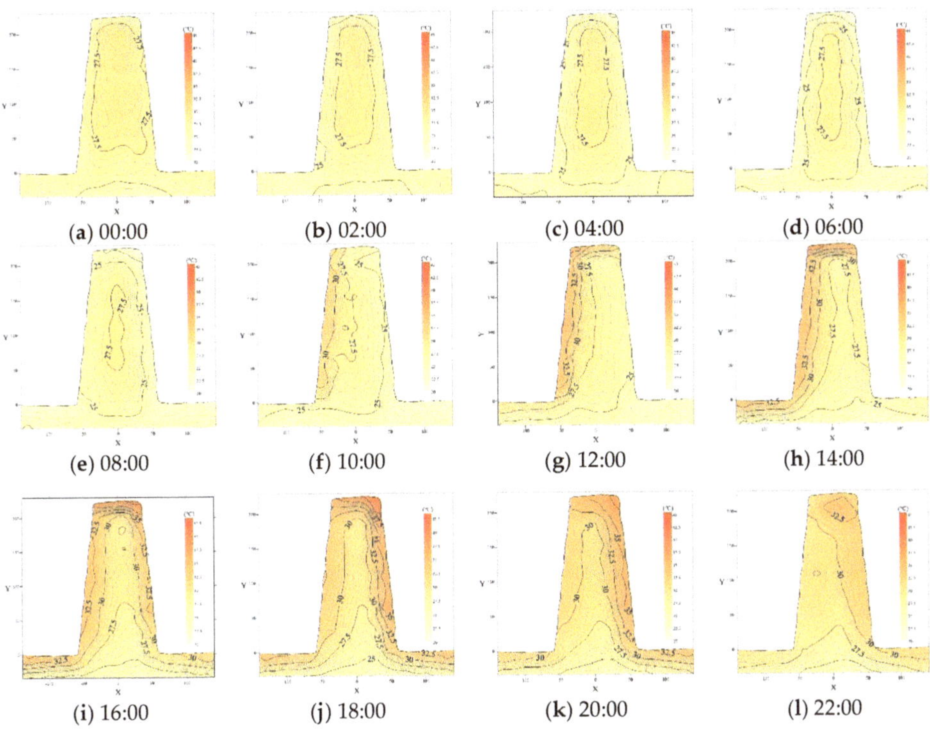

Figure 11. Temperature variation of test wall on June 16.

As can be seen from Figure 11, the internal temperature of the test wall was almost constant during this day in summer, and the largest temperature change was observed on the wall surface. The temperature gradually fell down to the lowest from 00:00 to 08:00, rose up to the highest from 08:00 to 16:00, and dropped slowly from 16:00 to 24:00. This indicates that the wall temperature gradually rose from 08:00 to 14:00, and the highest temperature was observed in the top corner area to the east of the test wall; the maximum temperature was 37.5 °C in this stage. The wall temperature was highest from 14:00 to 16:00, and the highest temperature area moved gradually from the east to the west, i.e., to the corner area on the west façade of the test wall, with a maximum temperature of 40 °C. The wall temperature decreased from 16:00 to 08:00 on the next day. It decreased slowly from 16:00 to 20:00 and then decreased sharply from 20:00 to 08:00 on the next day. The temperature curve gradually changed from an asymmetric to a symmetric elliptical spiral, and the surface temperature dropped to 22 °C.

According to the change in the summer temperature field, the wall was divided into four regions, as shown in Figure 12. The strong sensitive area is the red-filled part in the upper right corner at the top of the wall; this area exhibited the largest temperature change. Compared with the value at 08:00, the temperature had increased by 11.5 °C at 16:00. This area had a width of 25.54 cm and a height of 27.01 cm at the top west side of the wall. The sensitive area is the orange-filled area in Figure 12. It was distributed 15.58 cm from the top to the inside of the wall and 17.24 cm from the surface to the inside of the west side of the wall. Compared with the value at 08:00, the temperature had increased by 11.5 to 12.5 °C at 16:00. The transition area is the green-filled part in Figure 12. It was distributed around the wall as a parabolic shape, with a thickness of 15 cm. The maximum distance from the east side to the wall surface was 16.04 cm, the thickness at the top was 17.78 cm, and the distance from the foundation was 18.06 cm. Compared with the value at 08:00, the temperature had increased by 6 to 10 °C at 16:00. The stable area is the cyan-colored area in Figure 12. It was also distributed parabolically around the wall, with a thickness of 12.36 cm on the top, 19.66 cm on the east side, and 11.90 cm on the west side. Compared with the value at 08:00, the temperature increased by 4 °C at 16:00. The constant area is the blue-filled part; it exhibited almost no change in temperature, and its area increased by 2.65 times compared to that during spring.

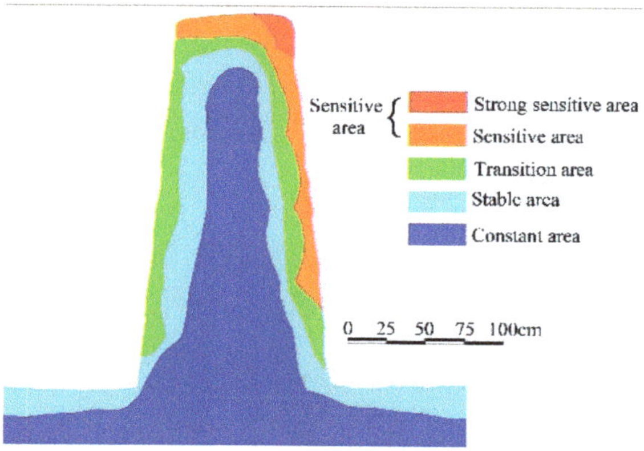

Figure 12. Temperature field of test wall in summer.

In summary, the summer temperature field was divided into four areas: the average temperature change in the sensitive area was above 10 °C; the temperature change in the transition area was 5–8 °C, and the stable area was 1–3 °C, which remained unchanged compared with the other time periods in the constant area, as seen in Figure 12.

4.2.3. Autumn

As can be seen from Figure 13, the sunshine was strongest from 00:00 on 22 September to 00:00 on 23 September. Starting from 00:00, data were collected every 4 h to complete the isotherm diagram, as shown in Figure 14.

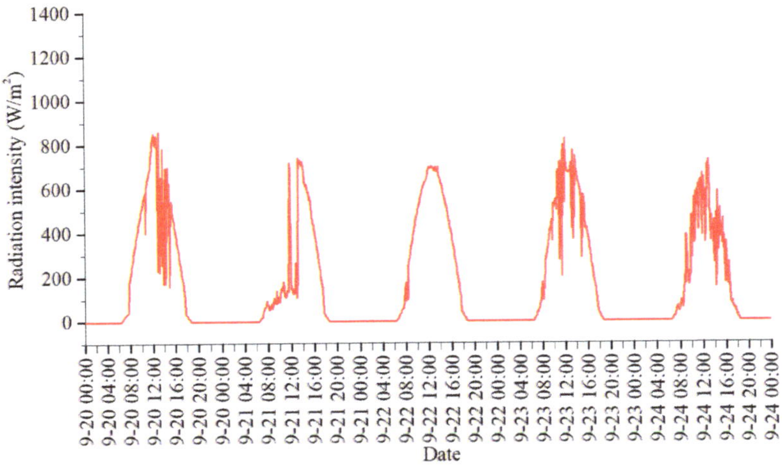

Figure 13. Selection of optimum days for autumn illumination.

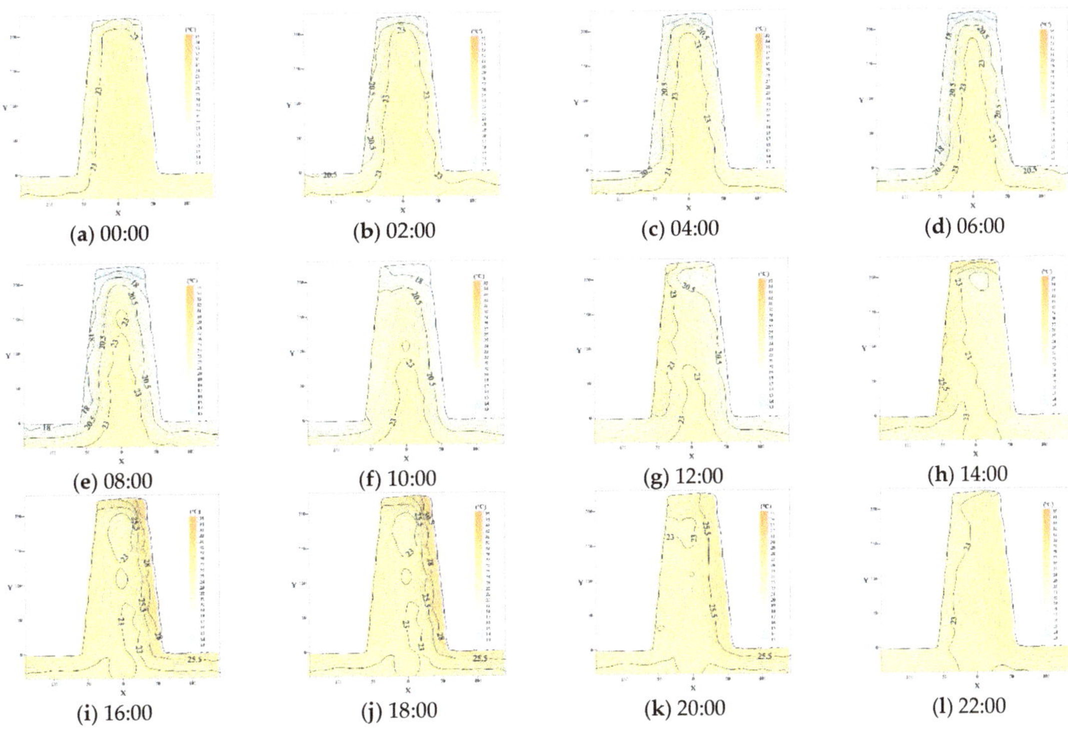

Figure 14. Temperature variations in the test wall on 22 September.

As can be seen from Figure 14, the internal temperature of the test wall was almost constant across one day in autumn, and the surface temperature changed the most. The temperature gradually fell down to the lowest from 00:00 to 08:00, rose up to the highest from 08:00 to 16:00, and dropped slowly from 16:00 to 24:00. This indicates that the wall temperature gradually rose from 08:00 to 14:00, and the highest temperature was observed in the top corner area to the east of the test wall; and the maximum temperature was 28 °C in this stage. The wall temperature was the highest from 14:00 to 16:00, and the highest temperature area moved gradually from the east to the corner area on the west facade of the test wall; it reached 40 °C. The wall temperature decreased from 16:00 to 08:00 on the next day; it decreased slowly before decreasing rapidly from 20:00 to 08:00. The decreasing trend gradually changed from an asymmetrical to a symmetrical parabola, and the surface temperature dropped to 16 °C.

According to the changes in the autumn temperature field, the wall was divided into four regions, as shown in Figure 15. The strong sensitive area is the red-filled part, with the largest temperature change. The wall temperature at 16:00 was 17 °C higher than at 08:00. This area was located 17.28 cm from the west wall and 34.07 cm from the top of the wall. The sensitive area is the red-filled part, which was located 32.14 cm away from the top and west edges of the wall. The maximum depth was 20.82 cm from the surface to the inside of the west wall, and it extended up to 16.09 cm away from the bottom of the wall. The wall temperature at 16:00 was 10–17 °C higher than at 08:00. The transition area is the green-filled part; it was distributed on the top and west of the wall. The thickness of the top layer was 15.66 cm, the west side was 17.26 cm thick, and the depth was 12.41 cm. The wall temperature at 16:00 was 10–17 °C higher than at 08:00. The constant area is the cyan-filled part. It was roughly triangular. The height of the constant region was 228.93 cm, its width was 44.75 cm, and height was slightly larger than that of the constant region in summer. The width was 1.2 times that of the constant region in summer, and the temperature remained almost unchanged in this area.

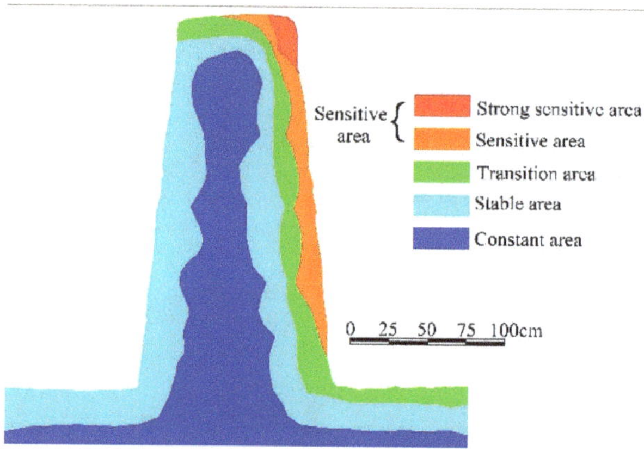

Figure 15. Temperature field of test wall in autumn.

To summarize, the autumn temperature field was divided into four areas, as shown in Figure 15; the average temperature change in the sensitive area was above 10 °C. The temperature change in the transition zone was between 5 and 8 °C, and the temperature change in the stable region was 1–3 °C. The temperature in the constant area did not change, compared with the other time periods in the constant region in Figure 15.

4.2.4. Winter

As can be seen from Figure 16, the sunshine was the brightest and most representative from 00:00 on 24 December to 00:00 on 25 December.

According to Figure 17, the internal temperature of the test wall is below zero. The internal temperature was almost constant, and the surface temperature did not change much across one day in winter. The temperature gradually fell down to the lowest from 00:00 to 08:00, rose up to the highest from 08:00 to 16:00, and dropped slowly from 16:00 to 24:00. The wall temperature gradually rose from 8:00 to 14:00, and the highest temperature was observed at the top corner area of the east side of the test wall (3 °C). The wall temperature was the highest from 14:00 to 16:00, and the highest temperature area was from the east to the west of the corner area on the west façade of the test wall (5 °C). The wall temperature decreased from 16:00 to 08:00 on the next day; it decreased slowly from 16:00 to 20:00, falling by 0.7–3 °C and then decreased rapidly from 20:00 to 08:00 on the next day. The curve gradually changed from a symmetric elliptic spiral to an asymmetric shape, and the surface temperature uniformly decreased to 6 °C.

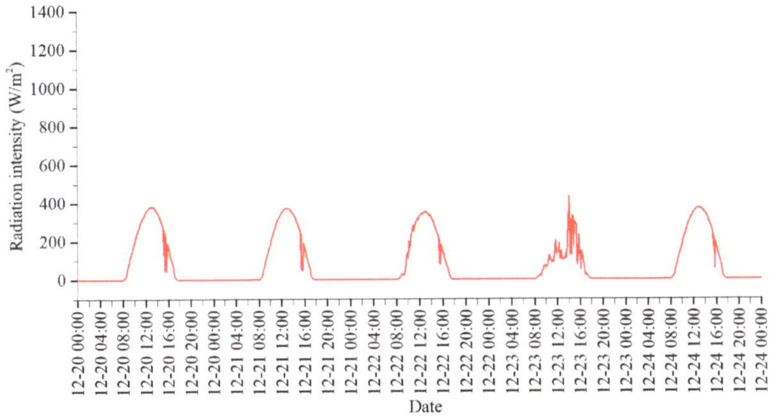

Figure 16. Selection of optimum days for winter illumination.

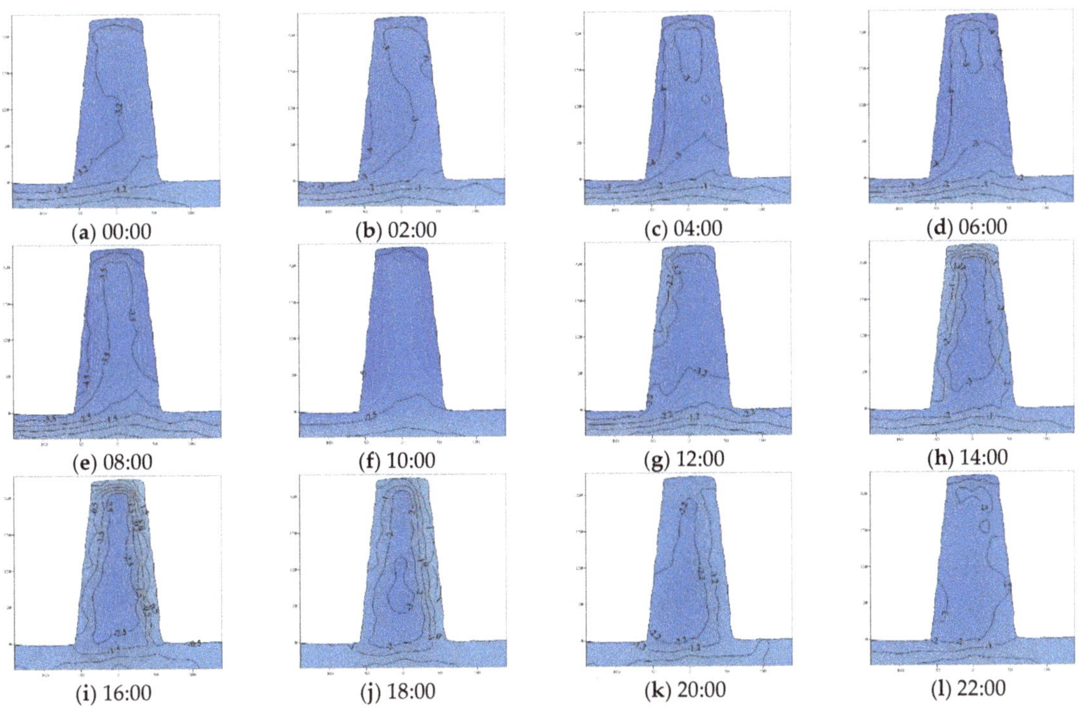

Figure 17. Temperature variations in the test wall on 22 December.

As can be seen from Figure 18, the overall temperature difference of the wall is less than 10 °C, so there was no sensitive area, only a transition area, a stable area and a constant area. The transition area is the green-filled part, which showed the largest temperature change. The wall temperature at 16:00 was −6 °C higher than it was at 08:00. The area was located 37.93 cm away from the west edge and 51.37 cm from the top of the wall. The constant area is the cyan-filled part in the middle and upper part on the east, west, and top of the wall. It was 13.66 cm thick on the top of the wall, 22.22 cm thick on the east side, and 107.28 cm deep from the foundation of the wall. The west side was at a depth of 20.18 cm. The constant area is the blue-filled part, which occupied 80.64% of the cross-sectional area of the entire wall. The temperature was almost unchanged in this range.

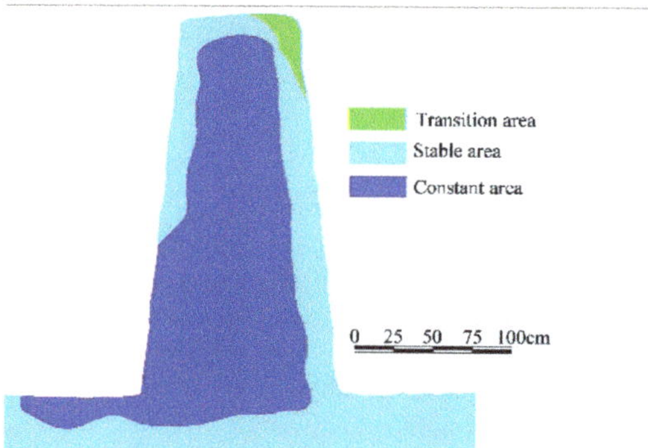

Figure 18. Temperature field of test wall in winter.

To summarize, the winter temperature field was divided into three areas (Figure 18), and the average temperature change in the transition area was 6 °C. The temperature change in the stable region was 2–3 °C. The third area, namely the constant area, showed no change in temperature compared with the other time periods Figure 18.

4.3. Variations in the Wall Temperature Field at Different Positions

The trend of the test wall temperatures and changes in the solar radiation angle meant that the test wall experienced different temperature variations during different seasons, at their different positions and depths, within one year. The temperature changes on the east and west sides of the wall within one year are shown in Figure 19. The temperature changes at the top, middle, lower, west, and east areas of the wall under the highest and lowest atmospheric temperatures are shown in Figure 20.

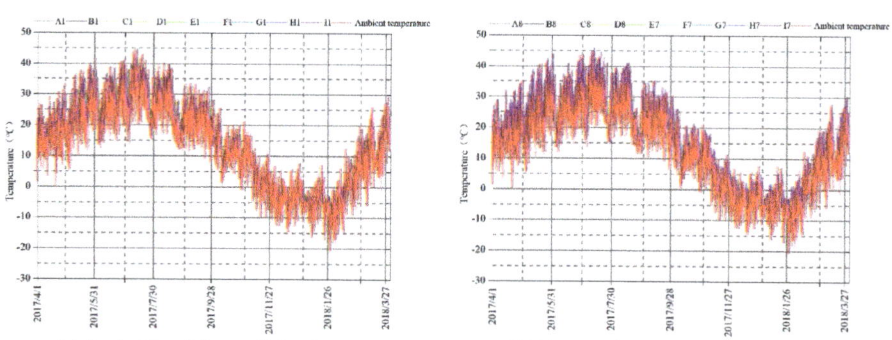

(**a**) East side of the wall in the year (**b**) West side of the wall in the year

Figure 19. Temperature change characteristic on the wall in the year (1 April 2017–31 March 2018).

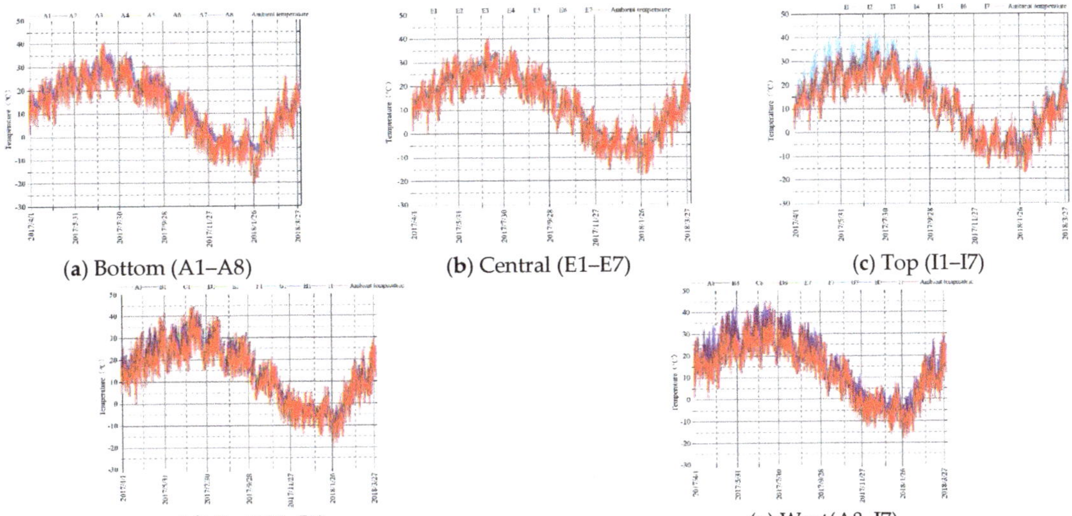

(**a**) Bottom (A1–A8) (**b**) Central (E1–E7) (**c**) Top (I1–I7)

(**d**) East(A1–I1) (**e**) West(A8–I7)

Figure 20. Wall temperature-atmospheric temperature relationship (1 April 2017–31 March 2018).

As can be seen from Figure 19, the temperature changes on the east and west sides of the wall were the same as those of the atmosphere, but the surface temperature of the wall changed rapidly. Within one year, the maximum temperature of the atmospheric environment was 41.24 °C, whereas the maximum temperature on the west side of the wall was 46.99 °C. The maximum temperature on the east side of the wall was 44.20 °C, which was 2.96 °C higher than the atmospheric temperature. The lowest atmospheric temperature was −20.5 °C. The lowest temperature on the west side of the wall was −15.2 °C, which was 5.3 °C higher than the atmospheric temperature, the lowest temperature on the east side of the wall was −15.4 °C, which was 5.1 °C higher than the atmospheric temperature. The maximum atmospheric temperature difference was 61.74 °C within one year, and the maximum temperature difference of the wall was 62.39 °C. The temperature of the east wall was not obviously warmer than the atmospheric temperature, while the temperature of the west wall was almost always higher than the atmospheric temperature, which was directly related to solar radiation.

As can be seen from Figures 20–26, the changes in temperature at the bottom, middle, top, east, and west of the wall showed the same trends, the changes in temperature at the bottom, middle, top, east, and west of the wall showed the same trends as that of the atmospheric temperature. However, the overall trend had a certain lag, which was no more than 6 h. As can be seen from Figure 20, there were severe wall temperature changes (from strong to weak) in the top, west, east, bottom, and middle areas. Along the vertical direction of the wall, the sensitivity of temperature to environmental influences, from strong to weak, ranged from the top to the middle and then to the bottom. Within one year, the highest temperature at the top of the wall was 46.99 °C, the lowest temperature was −15.3 °C, and the temperature difference was 62.29 °C. The highest temperature at the bottom of the wall was 37.3 °C, the lowest temperature was −5.9 °C, and the maximum temperature difference was 43.2 °C. The highest temperature in the middle was 35.8 °C, the lowest temperature was −5.6 °C, and the maximum temperature difference was 41.4 °C. Along the horizontal direction of the wall, the temperature change was more severe in the west than in the east, and it was relatively stable in the middle. The annual maximum temperature difference on the west side was 62.6 °C, the maximum temperature difference on the east side was 60 °C, and the maximum temperature difference in the middle was 36.8 °C.

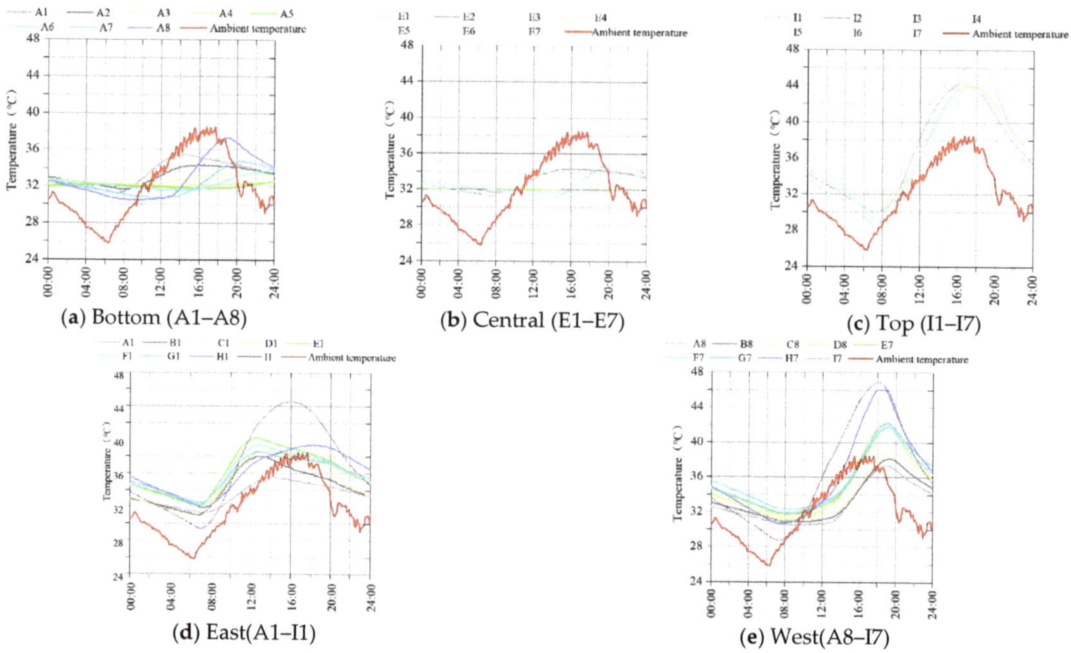

Figure 21. Variation of wall temperature during the highest atmospheric temperature throughout the year (13 July 2017).

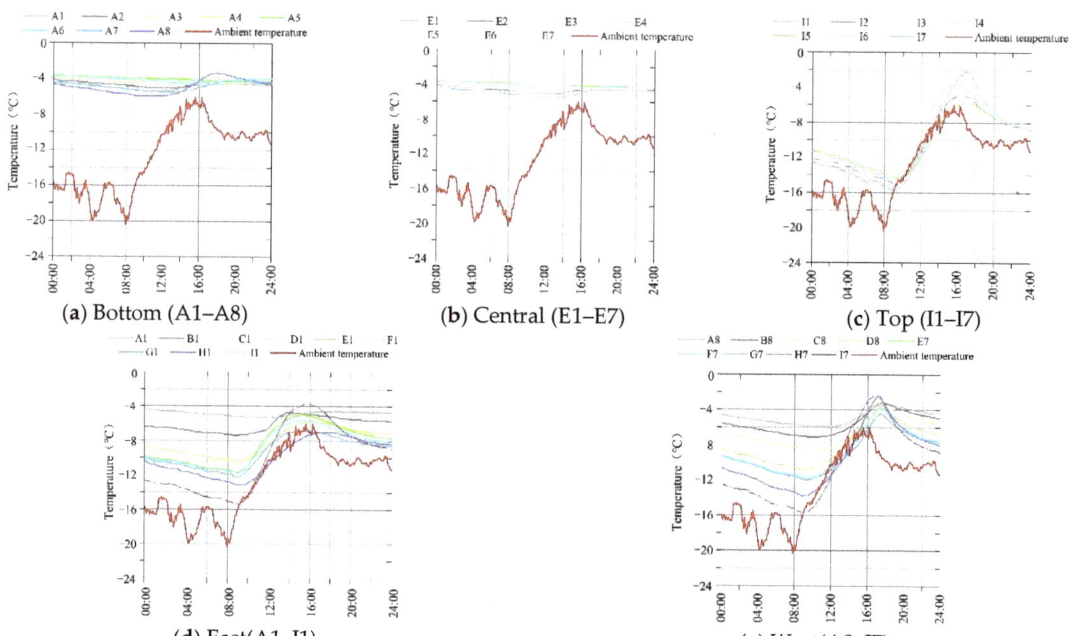

Figure 22. Variation of wall temperature during the lowest atmospheric temperature throughout the year (19 January 2018).

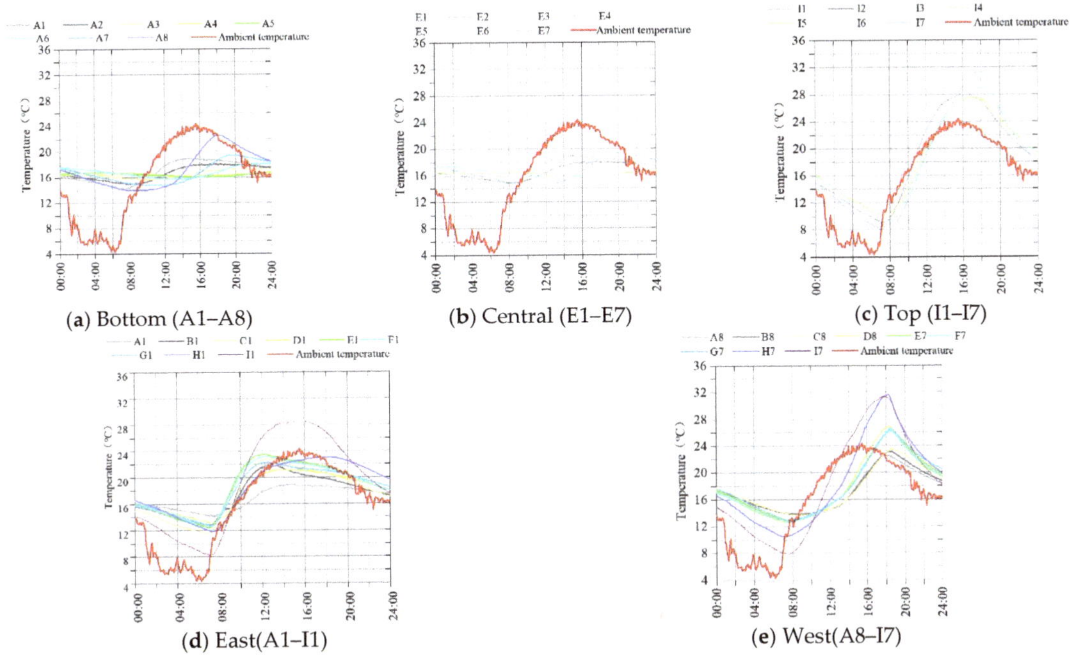

Figure 23. Diurnal wall temperature change with maximum atmospheric temperature difference in spring (22 April 2017).

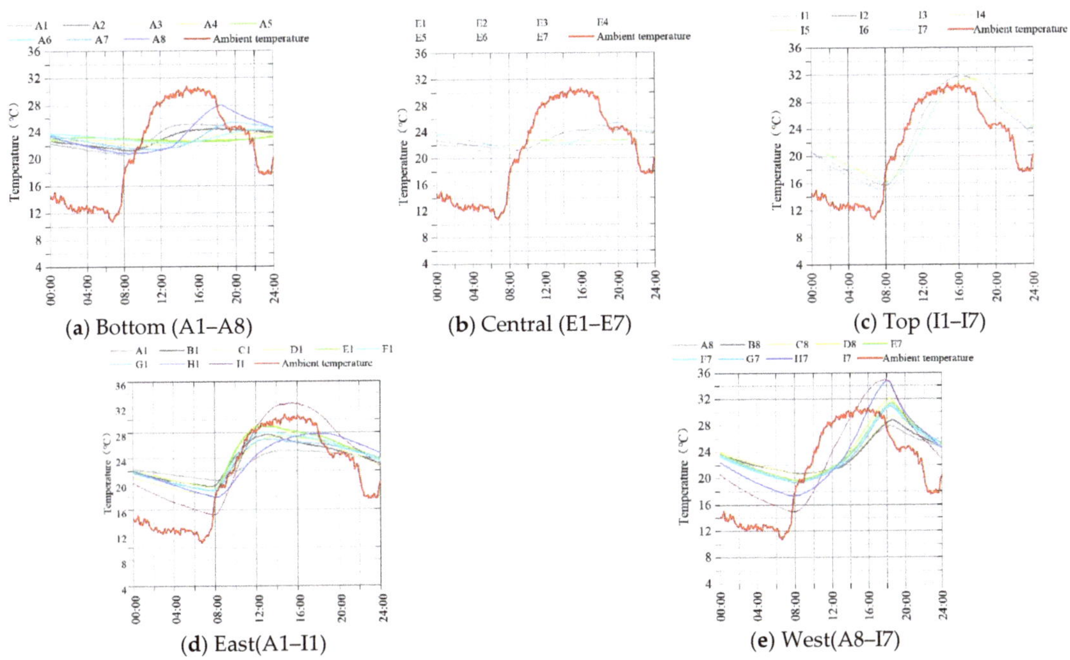

Figure 24. Diurnal wall temperature change with maximum atmospheric temperature difference in autumn (6 September 2017).

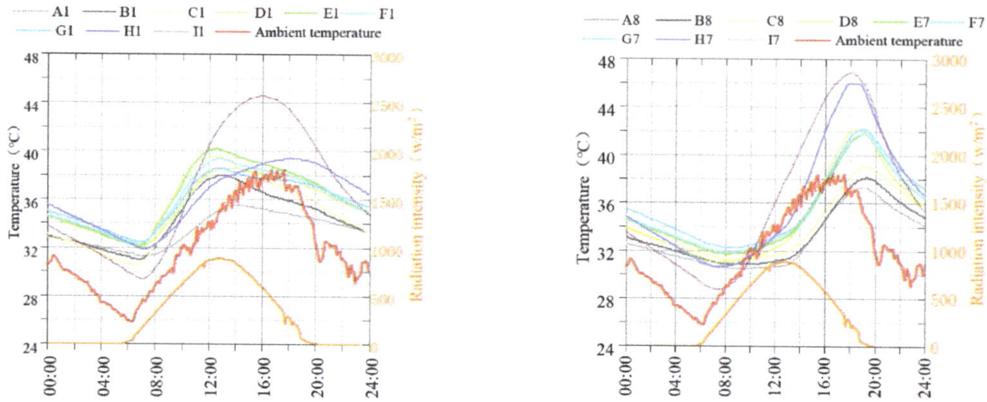

(**a**) East (A1–I1) (**b**) West (A8–I7)

Figure 25. The relationship of wall temperature, atmospheric temperature and the strongest solar radiation (13 July 2017).

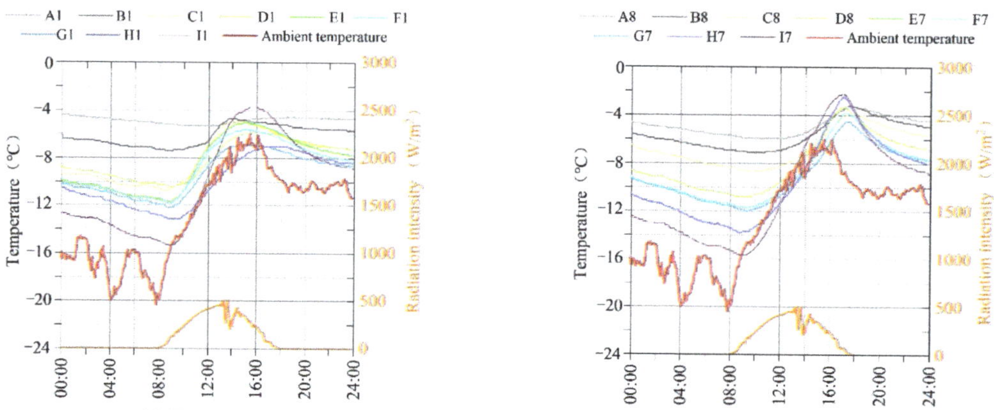

(**a**) East (A1–I1) (**b**) West (A8–I7)

Figure 26. The relationship of wall temperature, atmospheric temperature and the weakest solar radiation (19 January 2018).

As can be seen from Figure 21, on the hottest day of the year (13 July 2019), the maximum ambient temperature was 41.24 °C, and the wall temperature was 46.99 °C. The wall temperature showed a decreasing trend in the following order: top, west, east, environment, bottom, and middle. The surface temperature was higher than the internal temperature, with a larger overall gradient, and the maximum daily temperature difference was 18.6 °C. The temperature diffused from the arris of the west side to the top and lower parts of the wall. The west side temperature showed a decreasing trend in the following order: I7, H7, D8, G7, F7, E7, B8, and A8. The highest temperature was 46.99 °C, and the overall temperature rise showed a lag, with the same delay sequence. The temperatures of different test points on the east side from high to low were I1, H1, E1, C1, F1, G1, D8, B1, and A1. The highest temperature was 44.8 °C. The top temperature was the highest, the middle was in the second, and the bottom temperature was the lowest. In addition, the timing of the highest temperature at the top of the wall had a delay of about 1 h. The timing of the temperature rise in the other wall sensors was consistent with the ambient temperature, and the time at which the highest temperature was reached at the foundation was 2 h earlier than the ambient temperature. The order of the temperatures of the test points of the wall, from high to low, was I7, I6, I1, I2, I5, I3, and I4. The overall maximum temperature distribution was symmetrical in shape, and the temperature of the west side was higher than that of the east side and the middle. Furthermore, the temperature on the top layer of the east side was consistent with the maximum temperature of the environment. The time at which the highest temperature was reached on the west side of the wall was delayed by 2 h. The order of the temperatures of the test points in the middle of the test wall, from high to low, were E1, E2, E7, E3, E6, E4, and E5. The highest temperature on the east of the wall was higher than the west and middle of the wall, and the overall temperature was lower than the ambient temperature. The time at which the other test points reached their highest temperatures was delayed by 2–4 h in sequence. The order of temperatures at the bottoms of the test points, from high to low, was A8, A1, A7, A2, A6, A3, A5, and A4. The highest temperature on the west side of the bottom was higher than that on the east side, and the overall temperature was lower than the ambient temperature. The time it took to reach the highest temperature on the east side of the wall was 1–2 h lower than that under the ambient temperature, while the time taken to reach the highest temperature on the west side was delayed by 1–2 h.

As can be seen from Figure 22, the lowest ambient temperature occurred on 2018/1/19 (−20.5 °C); the lowest wall temperature was 15.4 °C. Compared with the ambient temperature, the order of the temperatures on each part of the wall, from high to low, was top, west side, east side, bottom, and middle. The surface temperature was lower than that of the interior and had a large overall gradient. The maximum daily temperature difference reached 13.7 °C, and the wall temperature diffused from the edges and corners on the west side of the wall to the top, middle, and lower parts. From low to high, the order of the temperatures of the west test points, from low to high, was I7, H7, G7, F7, E7, D8, C8, B8, and A8. The corresponding order for west test points was I7, H7, G7, F7, E7, D8, C8, B8, and A8. The overall trend gradually increased from top to bottom, which was higher than the lowest ambient temperature. The overall temperature decreased, and the time took to reach the lowest value lagged behind. The delayed time increased in reverse order. The daily temperature difference on the east side was smaller than that on the west side, and the maximum daily temperature difference was 11.3 °C. The order of the temperatures of the top test points, from low to high, was I7, I1, I2, I6, I5, I3, and I4. The overall maximum temperature was symmetrical in shape, and the lowest temperature in the west was lower than that in the east and lower than that in the middle; the top temperature was higher than the lowest temperature of the environment. The overall time taken to reach the lowest temperature was delayed by 1 h. There was no obvious change in the overall temperature of the middle part of the wall, and the overall temperature was higher than the highest ambient temperature. The maximum daily temperature difference was not more than 2 °C, and the order of the overall temperatures of the middle part, from low to high, was E7, E1, E2, E3, E6, E5, and E4. The overall temperature at the bottom of the wall showed no obvious change and was higher than the highest ambient temperature. The maximum daily temperature difference was not more than 3 °C. The order of the bottom temperature of the wall, from low to high, was A8, A1, A7, A2, A3, A6, A5, and A4.

As can be seen from Figures 23 and 24, on the day of the highest ambient temperature of the year, the maximum solar radiation reached 880 W/m^2 at 13:00. The sunshine lasted for more than 14 h on this day. In the morning, the sun shone on the east side of the wall, and the temperatures in the middle and lower parts rose to their highest at this moment. The temperature at the top still rose to its highest point at 16:00, in a certain range that was affected by solar radiation. The time at which the highest temperature was reached at the top of the west side was relatively delayed, reaching its peak at 18:00. The time at which the highest temperature was reached in the middle and upper parts was delayed by nearly 1 h, compared to the top. The bottom of the wall was the least exposed area to solar radiation, and compared to its lowest temperature, the top temperature on the west side rose by nearly 14.5 °C, and the top temperature on the east side rose by 12.6 °C. On the day of the lowest ambient temperature in a year, the solar radiation reached 500 W/m^2 at 14:00; the sunshine lasted for more than 10 h on this day. In the morning, the sun shone on the east side. At this time, the temperature rose to its highest in the middle and lower parts of the side wall. The temperature of the top, within a certain range, still increased under the influence of solar radiation, reaching its highest point slightly earlier than 16:00. The time at which the highest temperature was reached on the west side was relatively delayed (17:00). If the bottom of the wall is considered to be the least exposed to solar radiation and based on its lowest temperature, the top temperature on the west side rose by nearly 13.7 °C, and the top temperature on the west side rose by nearly 11.3 °C.

5. Discussion

Changes in atmospheric temperature and solar radiation directly affected the temperature field of the wall. As the buried temperature probes were nearly 5 cm away from the outer surface of the wall, the measured wall temperature was lower than the outer surface temperature of the wall. The wall temperature fluctuated with the fluctuations of atmospheric temperature and generally presented a certain lag. The temperature at the top and surface of the wall changed sharply and gradually decreased towards the center. The orders of the overall temperature in spring, summer, autumn, and winter, from high to low, were top, west, east, bottom, and middle. This order was entirely due to the influence of solar radiation.

It can be seen that the annual temperature difference of the wall was not lower than 36 °C, the maximum value was over 62.29 °C, the daily temperature difference was almost unchanged inside the wall, and the maximum daily temperature difference is 24.3 °C in the shallow surface layer of the wall, and the overall temperature of the wall gradually changed by 0.6 °C. Thus, drastic changes in the daily temperature difference of the shallow surface layer would be the main cause of the weathering of the wall surface. Particularly in spring and autumn, the wall surface temperature difference was large, and the wall surface temperature reached its highest value in the summer and the lowest in the winter. The temperature change of the wall was the main driving force for wall surface weathering. The daily temperature difference fluctuates around 0 °C in spring, which was the main external cause of wall deformation affected by the temperature field. In addition, the temperature difference gradient between the inner and outer layers of the wall was large, the temperature difference between different depth layers formed the difference between thermal expansion and cold contraction, and the difference in tensile and compressive stress between the inner layer interface was the external force of the surface exfoliation and pulverization. As can be seen from Figure 27, the temperature fluctuated around 0 °C for 101 days throughout the year. This mainly occurred in the winter (2017/11/10–2018/02/25), when the wall temperature circle was below 0 °C for 15 days. The wall's surface was equivalent to more than 100 freeze-thaw cycles, the inside of the wall was unthawed, and the outside had melted. The volume expansion and internal freeze shrinkage of the shallow layer was the main reason for the weathering and porosity of the shallow layer.

In addition, the sharp change in the daily temperatures at the wall surface, which is caused by heat conduction, heat convection, and heat radiation, is the main reason for wall surface weathering. As can be seen from Figure 28, the daily temperature difference of the atmospheric environment was greater than 20 °C for only 12 days in a year. The west side of the top of the wall was only at this level for 19 days, and the east side of the top was only for two days. The daily temperature difference was more than 10 °C for 286 days, the daily temperature difference on the west side of the top of the wall was more than 10 °C for 274 days, for the east side of the top of the wall this period was 234 days, for the west side of the middle it was only 55 days, for the east side it was 26 days, and there were no days warmer than 10 °C in the walls' foundation. The daily temperature difference was greater than 5 °C for 355 days, for the west side of the top of the wall this period was 349 days, for the east side of the top of the wall it was 346 days, for the west side of the middle it was 290 days, for the east side of the middle it was 257 days, for the west side at the bottom of the wall it was 118 days, for the east side of the bottom it was 13 days, and for the inside of the wall it was only one day. It can be seen that the days when the temperature difference was greater than 10 °C had the most prominent influence on the wall's surface. There were five times more days when the daily temperature difference was greater than 10 °C at the top of the wall than in the middle, and there were no days when the temperature difference was greater than 10 °C at the bottom. The overall wall temperature changed from 0 to 5 °C, and the overall daily temperature difference was greater than 2 °C. Temperature changes of less than 2 °C in the wall were caused by daily temperature differences. Therefore, a temperature change of less than 2 °C in the wall is the stable range of heat conduction.

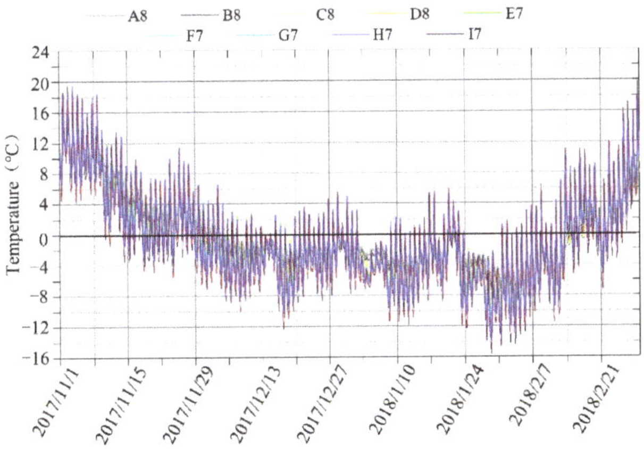

Figure 27. Fluctuations at 0 °C.

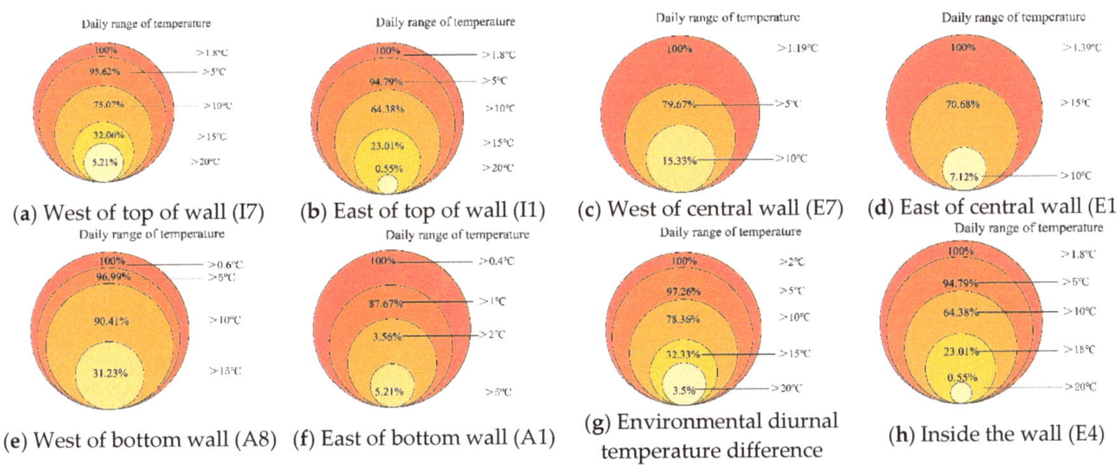

Figure 28. Statistical analysis of the daily temperature difference between the atmosphere and the wall.

The influence of ambient temperature changes on the temperature field of the test wall in spring, summer, autumn, and winter is shown in Figure 29. The temperature change on the west part of the wall was greater than that on the east in the sensitive sand transition areas. However, the east part showed greater change than the west part in the stable area. The surface temperature field changed most acutely in summer, followed by spring and autumn; it was relatively stable in winter. The thickness of the shallow surface layer in the temperature-sensitive area of the test wall was less than 32 cm, and the daily temperature change was more than 10 °C. The temperature change was severe (>30 °C) from 0 to an 18 cm depth of the wall surface. There was an obvious change in depths between 18 and 32 cm (not exceeding 32 cm).

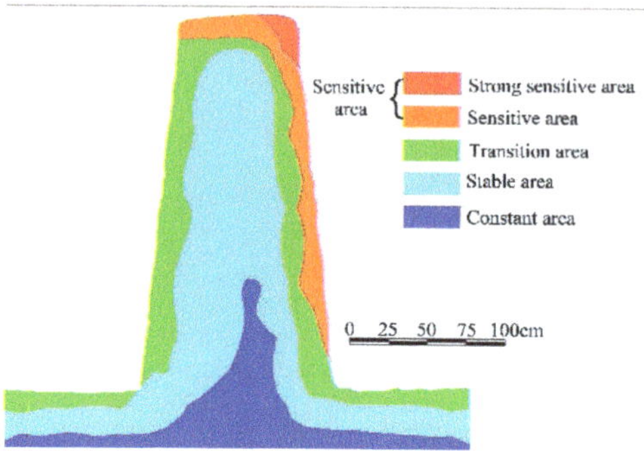

Figure 29. Temperature field of test wall across one year.

6. Conclusions

Heat conduction, convection, and radiation were identified as the main factors that caused temperature changes in rammed earth walls. The wall surface temperature changed sharply, while the interior remained relatively stable and showed a decreasing gradient, which was the external force for the weathering at the surface of the wall. The temperature field was responsible for the changes to the solid, liquid, and gaseous phases of water, which was the main mode of energy storage and release. Particularly in spring and winter, the alternating temperatures above and below 0 °C were more obviously responsible for the deterioration of the wall surface. The main conclusions were as follows:

(1) The order of the intensity of temperature changes at different positions on the wall, from high to low, was the top west, top east, top center, middle and upper part of the western wall surface, middle and upper part of the eastern wall surface, middle and lower part of the western wall surface; middle and lower part of the eastern wall surface, inner upper side, inner lower side, and finally internal middle. The daily temperature difference on the top surface of the wall was more than 10 °C, accounting for 75% of the days, and the temperature difference inside the wall did not exceed 2 °C. Temperature difference circulation was the main external cause of weathering on the wall surface.

(2) The maximum temperature difference on the surface of the wall was 62.99 °C. The daily gradual change of temperature was not more than 0.6 °C, and the maximum daily temperature difference was 24.3 °C. The changes were the most prominent in spring and autumn, with the highest temperatures occurring in summer and the lowest in winter. The alternating changes above and below 0 °C constituted another main reason for the repeated expansion and contraction weathering of the wall surface, and the extreme point of the annual temperature change was also an important factor.

(3) The temperature-sensitive area of the rammed earth wall was less than 30-cm deep. At 0–18 cm from the edge on the west side of the wall, the temperature change was sharp, whereas the daily temperature difference circulation exceeded 10 °C at depths deeper than 20 cm (the sensitive area). The transition area was 15 cm inside, and the daily temperature difference circulation at this place was 5–10 °C. Finally, in the stable area (10–30 cm inside), the daily temperature difference was 1–5 °C. The rest of the wall was termed as the constant area, whose temperature change was less than 1 °C.

(4) Solar radiation was the main factor that caused the sharp rises and observed lag in the wall temperature. The daily temperature difference between the environment

and the wall was the result of solar radiation. On the day with the highest ambient temperature during the year, the solar radiation reached 880 W/m^2 at 13:00 and the sunshine lasted for more than 14 h. On the day with the lowest ambient temperature during the year, the solar radiation was 500 W/m^2 at 14:00 and the sunshine lasted for more than 10 h. The daily temperature difference at the top of the west side of the wall was close, which fully showed that the daily temperature difference of the wall was closely related to the angle of solar radiation, the direction of the wall, and radiation time.

(5) The temperature difference in the daily cycle was the main factor that affected the wall deterioration. Over the duration of a year, the order of daily temperature difference, from high to low, was spring, autumn, summer, and winter. The temperature difference was the largest in spring and the most stable in winter. The temperature difference alternated above and below 0 °C in spring, and at this time the volume change because of thermal expansion and cold contraction of rammed soil was the largest. Therefore, spring constituted the main stage of temperature-induced deterioration of the rammed earth wall.

(6) The study revealed the temperature-sensitive areas of rammed earth walls affected by sunlight and environmental temperature, as well as the changes in different seasons, which accumulate experience for researching the deterioration mechanism of earthen sites under the multi-field coupling effect. It is expected to provide the basis for further developing the surface temperature field, stress field and coupling relationship of earthen sites, and provide the basis of environmental action parameters for studying the weathering mechanism of the superficial layer of earthen sites.

Author Contributions: Conceptualization, Q.P. and Q.G.; Data curation, B.Z. and J.Z.; Formal analysis, Q.P. and J.H.; Funding acquisition, Q.G.; Investigation, D.S.; Writing—original draft, Q.P.; Writing—review & editing, Q.G. All authors have read and agreed to the published version of the manuscript.

Funding: This work was supported by National key research and development plan "Study on Weathering Mechanism and Prevention and Control Technology of Multi-field Coupling Subsoil Sites" (Grant No. 2020YFC1522202); Talent Training and Introduction Program of CAS "Light of the West", Supported by Natural Science Foundation of Gansu Province, "On Dynamic Response Characteristics of Earthen Sites Reinforced by Ramming and Propping"(Grant No. 21JR7RA757).

Institutional Review Board Statement: Not applicable.

Informed Consent Statement: Not applicable.

Data Availability Statement: Not applicable.

Conflicts of Interest: The authors declare that they have no known competing financial interest or personal relationships that could have appeared to influence the work reported in this paper.

References

1. Soudani, L.; Woloszyn, M.; Fabbri, A.; Morel, J.C.; Grillet, A.C. Energy evaluation of rammed earth walls using long term in-situ measurements. *Sol. Energy* **2017**, *141*, 70–80. [CrossRef]
2. Serrano, S.; Barreneche, C.; Rincón, L.; Boer, D.; Cabeza, L.F. Optimization of three new compositions of stabilized rammed earth incorporating PCM: Thermal properties characterization and LCA. *Constr. Build. Mater.* **2013**, *47*, 872–878. [CrossRef]
3. Hall, M.R.; Allinson, D. Transient numerical and physical modelling of temperature profile evolution in stabilised rammed earth walls. *Appl. Therm. Eng. Des. Processes Equip. Econ.* **2010**, *30*, 433–441. [CrossRef]
4. Hall, M.; Allinson, D. Assessing the effects of soil grading on the moisture content-dependent thermal conductivity of stabilised rammed earth materials. *Appl. Therm. Eng.* **2008**, *29*, 740–747. [CrossRef]
5. Zhen, L.; Zhou, Z.; Zhang, H.; Sun, B.; Bai, L. Monitoring the temperature changes of earthen architecture walls. *J. Lanzhou Univ. (Nat. Sci.)* **2008**, *S1*, 58–60. (In Chinese)
6. Bui, T.T.; Bui, Q.B.; Limam, A.; Maximilien, S. Failure of rammed earth walls: From observations to quantifications. *Constr. Build. Mater.* **2014**, *51*, 295–302. [CrossRef]
7. Niroumand, H.; Zain, M.F.M.; Jamil, M. Various types of earth buildings. *Procedia-Soc. Behav. Sci.* **2013**, *89*, 226–230. [CrossRef]

8. Wang, X. New progresses on key technologies for the conservation of Chinese earthen sites in arid environment. *Dunhuang Res.* **2008**, *6*, 12–18. (In Chinese) [CrossRef]
9. Sun, M. A system for evaluating the deterioration of earthen sites. *Sci. Conserv. Archaeol.* **2012**, *24*, 27–32. (In Chinese) [CrossRef]
10. Sun, B.; Zhou, Z.; Zhang, H.; Zheng, L. The role of temperature in the weathering of rammed earth Sites. *Dunhuang Res.* **2009**, *6*, 72–76. (In Chinese) [CrossRef]
11. Richards, J.; Mayaud, J.; Zhan, H.; Wu, F.; Bailey, R.; Viles, H. Modelling the risk of deterioration at earthen heritage sites in drylands. *Earth Surf. Processes Landf.* **2020**, *45*, 2401–2416. [CrossRef]
12. Du, Y.; Chen, W.; Cui, K.; Zhang, J.; Chen, Z.; Zhang, Q. Damage Assessment of Earthen Sites of the Ming Great Wall in Qinghai Province: A Comparison between Support Vector Machine (SVM) and BP Neural Network. *ACM J. Comput. Cult. Herit.* **2020**, *13*, 1–18. [CrossRef]
13. Du, Y.; Chen, W.; Cui, K.; Zhang, K. Study on Damage Assessment of Earthen Sites of the Ming Great Wall in Qinghai Province Based on Fuzzy-AHP and AHP-TOPSIS. *Int. J. Archit. Herit.* **2020**, *14*, 903–916. [CrossRef]
14. Fodde, E.; Watanabe, K.; Fujii, Y. Measuring evaporation distribution of mud brick and rammed earth. *Struct. Surv.* **2014**, *32*, 32–48. [CrossRef]
15. Serrano, S.; de Gracia, A.; Cabeza, L.F. Adaptation of rammed earth to modern construction systems: Comparative study of thermal behavior under summer conditions. *Appl. Energy* **2016**, *175*, 180–188. [CrossRef]
16. Pei, Q.; Wang, X.; Guo, Q.; Zhang, B.; Zhao, G.; Zhao, J. Laboratory test of deformation mechanism of rammed roof-propping reinforcement at earthen heritage sites in arid environment. *Rock Soil Mech.* **2018**, *291*, 46–55. (In Chinese) [CrossRef]
17. Yang, S.; Wang, X.; Guo, Q.; Chen, Y.; Pei, Q. Experimental research on salt threshold of earthen sites in north west of China. *Lanzhou Univ. (Nat. Sci.)* **2018**, *54*, 5–12. (In Chinese) [CrossRef]
18. Wang, X.; Pei, Q.; Guo, Q.; Li, Z.; Wang, Y.; Zhao, J. Stress mechanism for the rammed layer interfaces of earthen heritage sites with different treatments. *J. Cult. Herit.* **2019**, *39*, 110–119. [CrossRef]
19. Shen, J.; Lassue, S.; Zalewski, L.; Huang, D. Numerical study on thermal behavior of classical or composite trombe solar walls. *Energy Build.* **2007**, *39*, 962–974. [CrossRef]
20. Taylor, P.; Luther, M.B. Evaluating rammed earth walls: A case study. *Sol. Energy* **2003**, *76*, 79–84. [CrossRef]
21. Soudani, L.; Fabbri, A.; Morel, J.C.; Woloszyn, M.; Chabriac, P.A.; Wong, H.; Grillet, A.C. Assessment of the validity of some common assumptions in hygrothermal modeling of earth based materials. *Energy Build.* **2016**, *116*, 498–511. [CrossRef]
22. Fernández, J.L.; Porta-Gándara, M.A.; Chargoy, N. Rapid on-site evaluation of thermal comfort through heat capacity in buildings. *Energy Build.* **2004**, *37*, 1205–1211. [CrossRef]
23. Ye, S. An interpretation of a game to the moon of the book of rites in terms of comparative mythology exemplified with the state of things in mid-spring. *J. Shanxi Norm. Univ. (Philos. Soc. Sci. Ed.)* **2006**, *35*, 5–10. (In Chinese) [CrossRef]
24. Liu, W.; Wang, L. Contrast ancient with modern about crop snowing date in Guanzhong included in the book of Fansheng-zhi. *XiandaiNongye Keji* **2010**, *18*, 20–21. (In Chinese) [CrossRef]
25. Zhou, Y. A folklore analysis of traditional geomantic culture. *J. Guangxi Univ. Natl.* **2006**, *26*, 123–127. (In Chinese) [CrossRef]
26. Chen, W.; Jia, B.; Cai, T.; Chen, H.; Li, X. Freeze-thaw deterioration of saline earthen sites under snowmelt or rainfall infiltration. *Chin. J. Geotech. Eng.* **2022**, *44*, 334–342. (In Chinese) [CrossRef]
27. Wu, P. *Study on the Damage Mechanical Characteristics of Artificially Prepared Site Soil by Dry-Wet and Freeze-Thaw*; Xi'an University of Technology: Xi'an, China, 2021. (In Chinese)
28. Chen, W.; Wei, D.; Lei, H.; Li, Y. Experimental study on strength deterioration characteristics of earthen sites covered by snow. *J. Lanzhou Univ. (Nat. Sci.)* **2019**, *55*, 655–660. (In Chinese) [CrossRef]
29. Lin, Z.; Tang, C.; Zeng, H.; Cheng, Q.; Tian, B.; Shi, B. Soil evaporation based on infrared thermal imaging technology. *Chin. J. Geotech. Eng.* **2021**, *43*, 743–750. (In Chinese) [CrossRef]
30. Zhang, H.; Yang, L.; Liu, P.; Chen, Y.; Zhang, G. Study on thermal deterioration simulation test of superficial layer on rammed earthen ruins. *J. Hunan Univ. (Nat. Sci.)* **2018**, *45*, 149–156. (In Chinese) [CrossRef]
31. Sun, B.; Zhou, Z.; Zhang, H.; Zhang, Y.; Zheng, L. Characteristics and prediction model of surface temperature for rammed earthen architecture ruins. *Rock Soil Mech.* **2011**, *3*, 867–871. (In Chinese) [CrossRef]
32. Zhang, B.; Wang, X.; Guo, Q.; Pei, Q.; Yang, S.; Li, F. Quality control research on reinforcement technology tests using rammed earth to fill unsupported walls in a Western Xia Imperial Tomb. *Dunhuang Res.* **2016**, *5*, 135–141. (In Chinese) [CrossRef]
33. Pu, T.; Chen, W.; Lü, H.; Du, Y. Analysis on function of deterioration of typical earthen ruins under the coupling of salinized and freezing and thawing in Qinghai–Tibet Plateau. *J. Cent. South Univ. (Sci. Technol.)* **2016**, *47*, 1420–1426. (In Chinese) [CrossRef]
34. Chen, M. *Intensity Deterioration Factor and Regulation's Study on Earthern Sites Undergoing Rainfall in Arid and Semi-Arid Regions*; Lanzhou University of Technology: Lanzhou, China, 2017. (In Chinese)

coatings

MDPI

Article

Properties and Compatibility Assessment of the Slurry with Quicklime for Fissure Restoration of Earthen Sites in Northwest China

Xiangpeng Yu [1], Kai Cui [1,*], Xue Hu [1], Qian Qiao [2] and Guojun Zhang [3]

[1] Western Center for Disaster Mitigation in Civil Engineering, Ministry of Education, Lanzhou University of Technology, Lanzhou 730050, China
[2] Department of Cultural Relics Management, Yinchuan 750002, China
[3] Northwest Research Institute Co., Ltd. of C.R.E.C., Lanzhou 730000, China
* Correspondence: cuik09@lut.edu.cn

Abstract: Slurries with quicklime for fissure restoration in earthen sites in northwest China can reduce dry shrinkage and reinforce the soil on both sides of fissure. However, their full-scale properties need to be investigated and a compatibility assessment needs to be carried out in order to proceed with further research. The paper pays attention to the full-scale properties of the slurry and typical site soils. A total of 3 slurries with different mix proportions and 35 earthen sites in northwest China are selected to test properties, including basic physical properties, mechanical properties, thermal properties, and water-physical properties. The test results show the properties of the slurries are close and that the properties of the site soils are obviously different. Then, a compatibility assessment is considered as a multi-criteria decision-making problem to solve. Three universal and necessary steps in the paper are to select assessment indicators, weigh each indicator, and sort all of the samples. The AHP, CRITIC, and TOPSIS methods are adopted in the process of the compatibility assessment. The results of assessment show that most site soils are commonly or highly compatible with three slurries; few site soils are very highly or lowly compatible; and that no site soils are very lowly compatible, which means that the three slurries with quicklime with different mix proportions are acceptable for most earthen sites in NW China. The paper provides much basic data of the slurries with quicklime and site soils widely distributed in arid region in NW China, and also builds a model of the compatibility assessment which has potential for the further application of slurries with quicklime.

Keywords: slurry with quicklime; earthen sites; property; compatibility assessment; multi-criteria decision-making

1. Introduction

Fissures are a typical and common erosion pattern of earthen sites in Northwest China, and fissure grouting is a common and effective method to repair fissures [1]. It has been suggested that soil can be used as repair materials for earthen buildings to improve their compatibility [2,3]. Typically used slurries in the field include slurry with calcined ginger [4], modified lime [5], and gypsum [6]. However, dry shrinkage of the soil gradually becomes an issue in conservation practice. Some researchers suggest that adding quicklime to the slurry may reduce shrinkage [7]. Fissure grouting on site shows that no new cracks develop, and that the soil on both sides of the fissure is strengthened due to the slurry expansion when adding quicklime [8]. Now, it is essential that full-scale properties be tested and analyzed for further potential applications. Thus, it means the compatibility assessment of the slurry with quicklime is crucial.

Compatibility is certainly one of the most frequently used words in conservation practice, and the concept of "more compatibility" is widely accepted by experts [9]. An essential component of conservation actions is the assessment of compatibility between conservation materials and relics. However, the compatibility requirements are not same for various conservation practice (for example, different relics [10–12] (lithoid, earthen, wooden, etc.), different climates [13,14] (humid, arid, etc.), and different purposes (restoration, consolidation, etc.). It is understood that a universal definition of compatibility to meet all requirements is not possible and that absolute compatibility is impractical. As mentioned in the literature [15]: " . . . *in spite of the fact that each one might have an idea of what are "proper and compatible" materials, it is felt that there is a need for a better definition of these concepts when historical structures are concerned, although bearing in mind that "absolute compatibility" is more a wishful thinking than an objective and feasible reality.*"

Although compatibility is not easy to define, some compatibility assessments are still often carried out. Unlike compatibility, incompatibility seems to be more easily approved by conservation workers. Some typical examples of incompatibility are that too much difference in strength leads to incompatible deformation, too much difference in color leads to a loss of historical information, and too much difference in gas/water permeability leads to interface separation [8], etc. This means that we can judge that the conservation material is incompatible with the exact relic when considering a certain character. This provides the possibility of operation for compatibility assessments [9].

The main goal of a compatibility assessment is to choose the most compatible materials or to optimize the conservation material for its specific application by varying several different parameters. So, the implication here may be that it is necessary to determine how to quantify the results of compatibility assessment and how to determine its relative position between "absolute compatibility" and "absolute incompatibility". These are the final outcomes that the compatibility assessment needs to present.

At present, qualitative results of compatibility assessment are common. Cui et al. [16] studied the mechanical compatibility of fissure grouting slurry with quicklime and thought that the slurry with the mass mix proportion of 3:2:5 and the liquid-solid ratio of 0.5 is the most compatible due to the most similar mechanical parameters. Loureiro et al. [17] studied the assessment of compatibility between historic mortars and lime-METAKAOLIN restoration mortars made from amazon industrial waste, and the conclusion is the restoration mortars are compatible with historic mortars as the mineralogical composition and physical and mechanical properties are similar. Saez-Perez et al. [18] studied the characterization and chromatic assessment of gypsum-based pastes for construction and heritage restoration, and thought the colorimetric analysis by means of quantitative spectrophotometry is of vital importance when determining the loss of color intensity of the pastes used. Lubelli et al. [19] studied the characterization and compatibility assessment of commercial stone repair mortars, and the conclusion is some requirements are hard to be fulfilled and not all requirements can be fulfilled at the same time. Sala et al. [20] studied the lightweight natural lime composites for rehabilitation of historical heritage, and thought several natural lime composites satisfies different strength/density requirements. Mazzuca et al. [21] studied polyvinyl alcohol-based hydrogels as new tunable materials for application in the cultural heritage field, and thought the reported gels are optimum candidates for paper cleaning applications due to appropriate mechanical and retentive properties. Bal et al. [22] studied the oxidation-resistant copper nanoparticles and their possible application in nanofluid and nanohydrogels, and the nanoparticles were prepared with biocompatible and naturally available polymers.

Meanwhile, according to the related researches, the compatibility includes multiple branches, such as physical-chemical branch, mechanical branch, artistic branch, and so on, even though a particular conservation is concerned. Obviously, compatibility assessment becomes a multi-criteria decision-making (MCDM) problem when considering every branch of compatibility. Lots of studies have been reported for MCDM in different fields and some typical methods are widely adopted [23–26].

In this paper, the focus is the properties and assessment of compatibility between the slurry for fissure grouting and site soils in Northwest China. The four branches of properties (basic physical, mechanical, thermal, water-physical) are tested whether the slurry and site soils. Based the results of properties, the compatibility assessment is carried out. Total 12 indicators are selected, the combined weight is produced by the subjective and objective weights, and all site soils are ranked with TOPSIS. The research may have an obvious impact on some practical applications.

2. Materials and Methods

2.1. Materials

2.1.1. Slurries

According to the previous related researches [7,8], a micro-expansion is caused when the mass proportion of quicklime reaches 30%. To ensure the appearance of the micro-expansion, the mass proportion of quicklime adopts 30%, and the mass proportion of fly ash adopts 10%, 20% and 30%. To keep a similar consistency, corresponding with the mass proportion of fly ash, the liquid-solid ratio adopts 0.52, 0.50, and 0.48 [7]. Table 1 shows the details of mix proportions (mass) of slurries.

Table 1. Details of mix proportions (mass) of slurries.

Abbreviation	Quicklime	Fly Ash	Soil	Liquid-Solid Ratio
3:1:6	30%	10%	60%	0.52
3:2:5	30%	20%	50%	0.50
3:3:4	30%	30%	40%	0.48

Quicklime is purchased from Tianyu Fine Chemical Company in Lanzhou, China. Based on laboratory test results, the standard designation of the quicklime is EN 459-1 CL 90-Q (R5, P1) in accordance with Building Lime—Part 1: definitions, specifications and conformity criteria (BS EN 459-1: 2015).

Fly ash is purchased from Xigu Thermal Power Plant, Lanzhou, China. The chemical composition is shown in Table 2. The fly ash belongs to Category N in accordance with Fly ash for concrete Part 1: Definition, specifications and conformity criteria (BS EN 450-1: 2012). Meanwhile, fly ash is original and has not undergone any alkaline activation process.

Table 2. Chemical composition of fly ash.

Mass	SiO_2	Al_2O_3	Fe_2O_3	CaO	MgO	SO_3	LOI
%	55.9	27.4	5.5	4.9	1.5	1.4	1.1

Soil comes from the Gouya Mountain, Lanzhou, China, and its physical properties are shown in Table 3.

Table 3. Physical properties of soil.

Name	Density ($g{\cdot}cm^{-3}$)	Water Content (%)	Liquid Limit (%)	Plastic Limit (%)	Specific Gravity
Silty clay	1.66	0.86	34.49	21.02	2.61

Modified polyvinyl alcohol solution (often called *SH* solution in Chinese literature) is colorless and odorless, and can be diluted indefinitely by water. The raw concentration of modified polyvinyl alcohol solution is 5% (mass fraction), and its density is 1.09 g/cm^3 and its viscosity is 51.45×10^{-3} Pa·s at a temperature of 25 °C. The concentration of 1.5% is used in this work. The purpose of adding modified polyvinyl alcohol (PVA) solution is to improve the cohesion and the strength of the slurry since PVA molecules form the film covering and the filiform bonding around soil particles, and eventually connect to form a stable spatial mesh network [27,28]. Note that modified polyvinyl alcohol solution can stay stable indoors. A solid film can only form in the wall of the container when the solution in the container dries. A solid jelly can form when the solution is exposed in low temperature (below 0 °C) for a long time. The film and the jelly are no longer soluble in water.

The following is the manufacture process of the slurry. First, the mixture of quicklime, fly ash and site soil with the designed mix proportion is stirred at a speed of 450 r/min for two minutes. Then, the binder is mixed and stirred at the same speed for one minute according to the designed liquid-solid ratio. Finally, the prepared slurry is poured into the mold to shape. The slurry is shaped for 24 h in the cube model of 70.7 mm. The slurry sample is exposed to the indoor air for 63 d. All specimens for tests are cut from the original cube sample.

2.1.2. Site Soils

All site soils are picked out form the crumbling part of each earthen ruins (Figure 1a), and then carefully wrapped in tissue paper and packed in sticky tape (Figure 1b). Site soils are transported into the laboratory, and then a series of tests are taken in the laboratory.

(a) (b)

Figure 1. Crumbling earthen ruins and wrapped and packed soils. (**a**) Crumbling earthen ruins; (**b**) wrapped and packed soils.

Thirty-five typical earthen sites distributed in Northwest China are adopted. Table 4 shows the basic information of every earthen site, such as the latitude, the longitude, the architectural form, and the construction period. Furthermore, codes ES01~ES35 indicate the earthen site from north to south.

Table 4. Basic information of 35 earthen sites.

Code	Location	Latitude (N)	Longitude (E)	Construction Dynasty	Architectural Form
ES01	Guazhou	40°29′37″	95°32′38″	Han	Beacon tower
ES02	Guazhou	40°25′38″	95°41′35″	Qing	Castle
ES03	Guazhou	40°15′34″	95°35′54″	Tang	Castle
ES04	Dunhuang	40°14′22″	94°47′21″	Song	Castle
ES05	Jiayuguan	39°54′54″	98°23′38″	Ming	Great Wall
ES06	Jiuquan	39°51′13″	98°34′33″	Ming	Great Wall
ES07	Gaotai	39°29′03″	99°43′06″	Ming	Great Wall
ES08	Linze	39°12′00″	100°22′28″	Ming	Great Wall
ES09	Zhangye	38°51′53″	100°45′46″	Ming	Great Wall
ES10	Minqin	38°40′00″	103°0′59″	Qing	Castle
ES11	Shandan	38°39′52″	101°15′13″	Ming	Great Wall

Table 4. *Cont.*

Code	Location	Latitude (N)	Longitude (E)	Construction Dynasty	Architectural Form
ES12	Ningxia	38°36.925′	106°4.297′	Ming	Castle
ES13	Minqin	38°33′19″	103°00′20″	Han	Castle
ES14	Ningxia	38°33.33′	105°83′51′	Ming	Great Wall
ES15	Jinchang	38°30′18″	102°25′30″	Tang	Castle
ES16	Minle	38°27′37″	100°52′27″	Ming	Great Wall
ES17	Yongchang	38°19′34″	101°56′38″	Ming	Great Wall
ES18	Jinchang	38°09′56″	102°08′32″	Song	Castle
ES19	Wuwei	37°41′59″	102°58′44″	Ming	Great Wall
ES20	Gulang	37°28′02″	102°54′32″	Ming	Great Wall
ES21	Menyuan	37°22′42″	101°37′05″	Ming	Great Wall
ES22	Jingtai	37°09′52″	104°10′20″	Ming	Beacon tower
ES23	Tianzhu	37°08′51″	102°52′25″	Ming	Castle
ES24	Huzhu	36°45′55″	101°56′13″	Ming	Great Wall
ES25	Baiyin	36°44′51″	104°41′53″	Tang	Castle
ES26	Yongdeng	36°42′45″	103°16′15″	Qing	Castle
ES27	Yongdeng	36°40′41″	102°53′55″	Song	Castle
ES28	Ledu	36°29′03″	102°23′57″	Ming	Great Wall
ES29	Yongdeng	36°27′07″	103°22′45″	Han	Castle
ES30	Pingan	36°27′02″	102°04′02″	Ming	Great Wall
ES31	Huangzhong	36°25′38″	101°35′25″	Ming	Great Wall
ES32	Minhe	36°19′49″	102°51′47″	Ming	Great Wall
ES33	Yongdeng	36°18′53″	103°24′58″	Ming	Great Wall
ES34	Yongjing	36°04′50″	103°17′02″	Ming	Great Wall
ES35	Guide	36°02′54″	101°25′57″	Ming	Great Wall

2.2. Tests

The same tests are taken for three slurries and thirty-five site soils in the laboratory, and four branches of the properties are considered in the paper. Three replicates samples are prepared for each test.

For basic physical properties, the density, specific gravity, and water content are tested meeting the *Standard for geotechnical testing method (GB/T 50123-2019)*. The density is tested with the wax sealed method, the specific gravity is tested with the pycnometer and the water content is tested with the drying method.

For mechanical properties, the compression and tension strength, inner cohesion and internal friction angle are tested. The compression test is uniaxial and unconfined, and the tension test is with the Brazilian tensile method. The loading machine is the electro-hydraulic press machine with the maximum load of 30 kN (Figure 2a), and the loading rate is 1 mm/min for the compression test and 0.5 mm/min for the tension test. From the load-displacement curve, the stress-strain curve can be calculated. The modulus of elasticity is the slope of the linear part in the stress-strain curve. The inner cohesion and internal friction angle are obtained through the direct shear test. For the soil samples (φ61.8 mm × 20 mm), the normal stress (σ) is designed as 100, 200, 300, and 400 kPa, and the shear rate is designed as 0.8 mm/min. The maximum of the shear stress (τ) is recorded corresponding with the normal stress, and then the four point is drawn and fitted with the line in the σ-τ coordinate system. The equation of the fitting line is certain, and the constant is the inner cohesion and the slope is the tangent of the internal friction angle according to the Mohr-Coulomb law. The test machine is the ZJ strain-controlled shear machine from Nanjing Soil Instrument Factory (Figure 2b).

For thermal properties, the thermal conductivity, coefficient of thermal diffusion and coefficient of heat expansion are tested. The thermal tests are taken using Xiangke Test Machine for Thermal Conduction DRE-2C. The thermal conductivity and the coefficient of heat diffusion is tested by Transient plane source method (TPS) meeting the standard *Plastics—Determination of the thermal conductivity and thermal diffusivity- Part 2:*

Transient plane heat source (hot disc) method (ISO22007-2-2008), and the size of the specimen is 5 cm × 5 cm × 1.5 cm. The specific heat capacity is calculated with the density, thermal conductivity and coefficient of thermal diffusion. The coefficient of heat expansion is tested with Dongfang Test Machine for heat expansion PCY-III-1400, and the size of the specimen is φ8 mm × 23 mm. The rising rate of temperature is 2 °C/min and the maximum of temperature is 150 °C in the test process.

(a) (b)

Figure 2. The machine for mechanical tests. (**a**) Loading machine; (**b**) shear machine.

For water-physical properties, the permeability coefficient is tested according to the Chinese *Standard for geotechnical testing method (GB/T 50123-2019)* and the disintegration rate is tested according to the literature [29,30].

3. Results

3.1. Test Results of Slurries

3.1.1. Basic Physic Test

Figure 3 shows the results of basic physic test, including density, specific gravity, water content, and porosity. The slurry with the mix proportion 3:2:5 has the minimum of the density, specific gravity and water content, which shows there are more pores in the slurry. So, the porosity of the slurry with 3:2:5 is the maximum among three slurries. The specific gravity decreases with the increase of the proportion of fly ash, other indexes do not. On the whole, the basic physical properties of the three slurries are close.

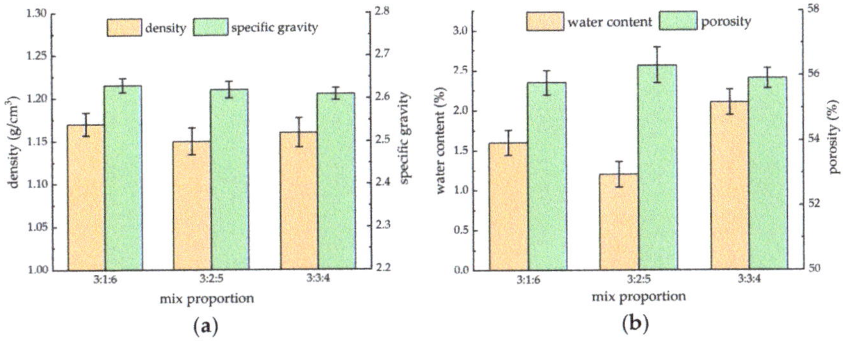

Figure 3. Basic physical properties of three slurries. (**a**) Density and specific gravity; (**b**) water content and porosity.

3.1.2. Mechanical Test

As can be seen from Figure 4a, the compression strength of slurries with mix proportion 3:1:6, 3:2:5, and 3:3:4 is 1.48, 1.575, and 1.524 MPa; the tensile strength is 0.229, 0.243, and 0.235 MPa. As can be seen from Figure 4b, the modulus of elasticity of slurries with mix proportion 3:1:6, 3:2:5, and 3:3:4 is 180, 202, and 193 MPa. As can be seen from Figure 4c, the inner cohesion of slurries with mix proportion 3:1:6, 3:2:5, and 3:3:4 is 232, 241, and 215 kPa; the internal friction angle is 39.65°, 43.92°, and 40.12°. The slurry with 3:2:5 has the maximum of the all mechanical properties. It is obvious that the mechanical properties of the slurry with mix proportion 3:2:5 are the best.

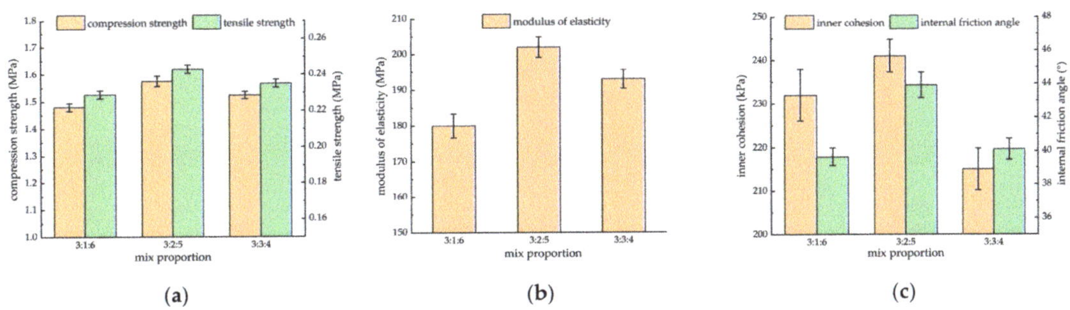

Figure 4. Mechanical properties of three slurries. (**a**) Compression and tensile strength; (**b**) modulus of elasticity; (**c**) inner cohesion and internal friction angle.

3.1.3. Thermal Test

Figure 5a,b show the results of thermal test for slurries. The thermal conductivity of slurries with mix proportion 3:1:6, 3:2:5, and 3:3:4 is 0.809, 0.825, and 0.814 W/m/K; the coefficient of thermal diffusion is 0.809, 0.953, and 0.809 m^2/s. The specific heat capacity of slurries with mix proportion 3:1:6, 3:2:5 and 3:3:4 is 0.855, 0.753, and 0.867 kJ/kg/K; the coefficient of heat expansion is 5.68×10^{-6}, 5.35×10^{-6}, and 5.51×10^{-6} 1/K. Overall, high thermal conductivity and coefficient of thermal diffusion and low specific heat capacity of the slurry with mix proportion 3:2:5 could reduce heat accumulation, and low coefficient of heat expansion could reduce the heat deformation.

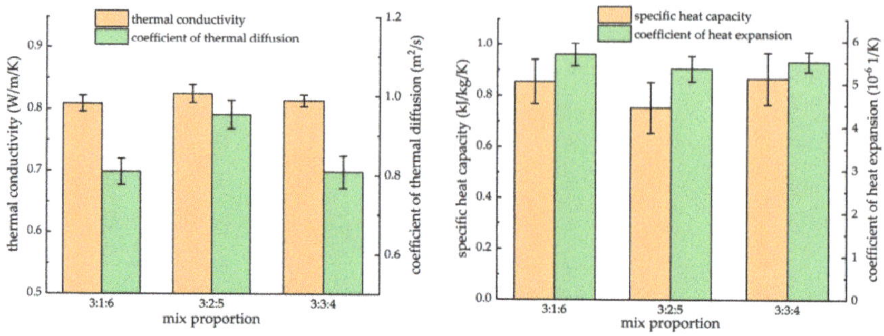

(a) (b)

Figure 5. Thermal properties of three slurries. (**a**) Thermal conductivity and coefficient of thermal diffusion; (**b**) specific heat capacity and coefficient of heat expansion.

3.1.4. Water-Physical Test

As can be seen from Figure 6, the permeability coefficient of slurries with mix proportion 3:1:6, 3:2:5 and 3:3:4 is 18.7×10^{-6}, 23.2×10^{-6} and 21.3×10^{-6} cm/s. The slurry with 3:2:5 has the maximum of the permeability coefficient. After being soaked for enough time, the mass of all slurries does not change and the disintegration does not occur. So, the disintegration rate of three slurries is 0 g/min.

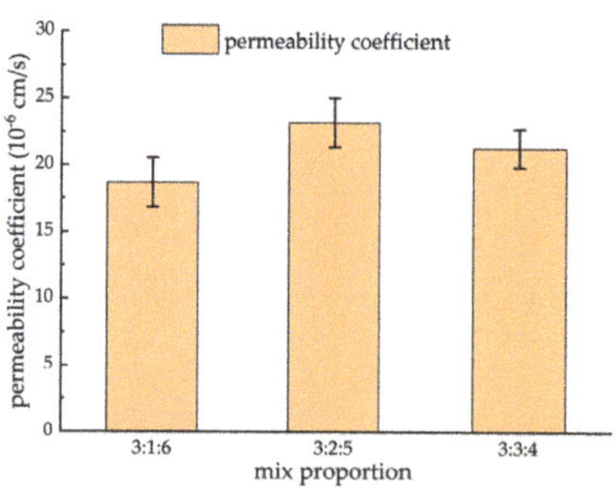

Figure 6. Water physical properties of three slurries.

3.2. Test Result of Site Soils

3.2.1. Basic Physic Test

As can be seen from Figure 7, the maximum and minimum of the density of 35 site soils is 2.11 g/cm^3 and 1.38 g/cm^3, and the average (marked with the red dashed line and the red arrow and text in Figure 7a) is 1.633 g/cm^3; the maximum and minimum of the specific gravity of 35 site soils is 2.74 and 2.47, and the average (marked with the red dashed line and the red arrow and text in Figure 7b) is 2.614; the maximum and minimum of the water content of 35 site soils is 4.07% and 0.12%, and the average (marked with the red dashed line and the red arrow and text in Figure 7c) is 1.604%; the maximum and minimum of the density of 35 site soils is 47.87% and 19.92%, and the average (marked with the red dashed line and the red arrow and text in Figure 7d) is 38.478%; the maximum and minimum of the plasticity index of 35 site soils is 12.57 and 8.28, and the average (marked with the red dashed line and the red arrow and text in Figure 7e) is 10.153; the maximum and minimum of the liquidity index of 35 site soils is −0.93 and −2.84, and the average (marked with the red dashed line and the red arrow and text in Figure 7f) is −1.830.

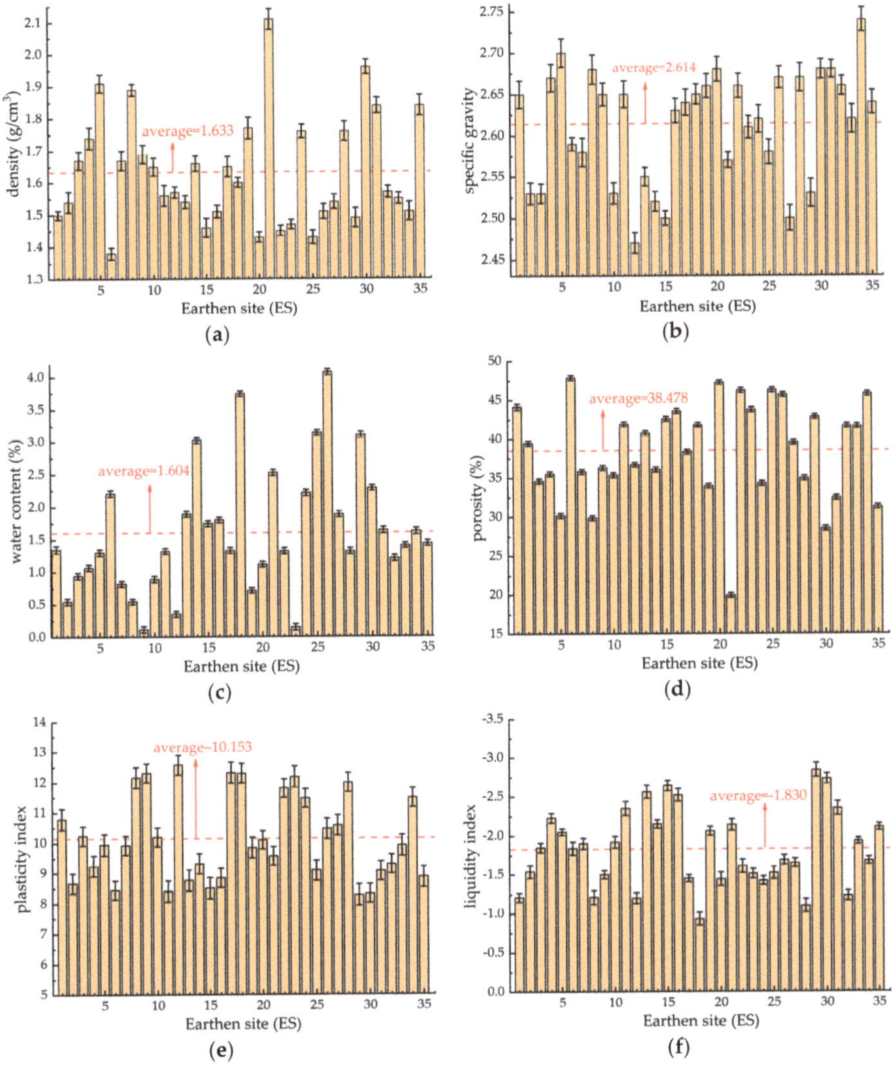

Figure 7. Basic physical properties of 35 site soils. (**a**) Density; (**b**) specific gravity; (**c**) water content; (**d**) porosity; (**e**) plasticity index; (**f**) liquidity index.

3.2.2. Mechanical Test

As can be seen from Figure 8, the maximum and minimum of the compression strength of 35 site soils is 2.303 and 0.146 MPa, and the average (marked with the red dashed line and the red arrow and text in Figure 8a) is 0.859 MPa; the maximum and minimum of the tensile strength of 35 site soils is 0.35 and 0.01 MPa, and the average (marked with the red dashed line and the red arrow and text in Figure 8b is 0.118 MPa; the maximum and minimum of the modulus of elasticity of 35 site soils is 212 MPa and 11 MPa, and the average (marked with the red dashed line and the red arrow and text in Figure 8c) is 76.17 MPa; the maximum and minimum of the inner cohesion of 35 site soils is 393 and 3 kPa, and the average (marked with the red dashed line and the red arrow and text in Figure 8d) is 112.14 kPa; the maximum and minimum of the internal friction angel of 35 site soils is 54.4° and 18°, and the average (marked with the red dashed line and the red arrow and text in Figure 8e) is 35.81°.

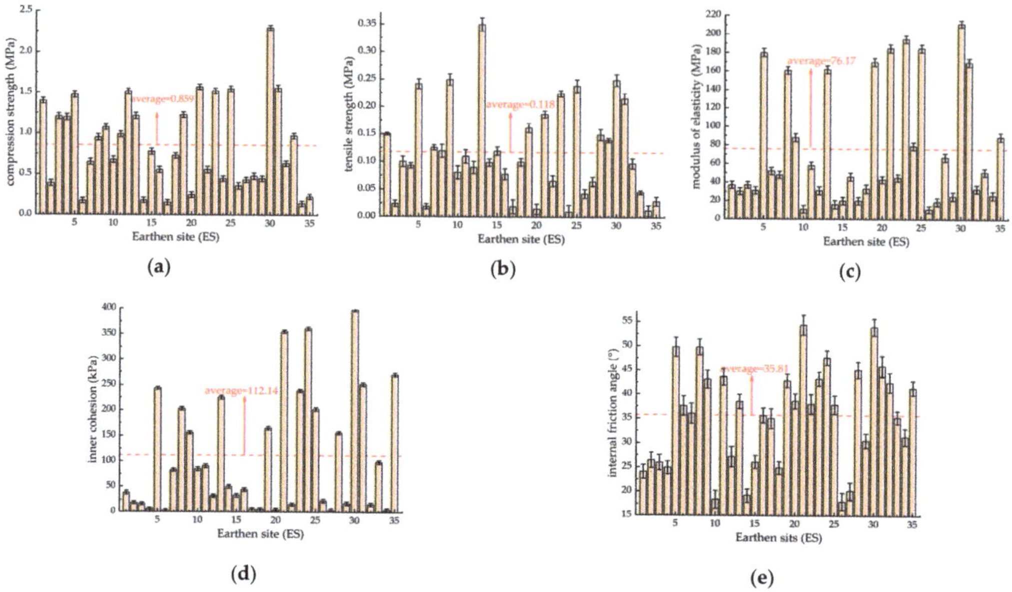

Figure 8. Mechanical properties of 35 site soils. (**a**) Compression strength; (**b**) tensile strength; (**c**) modulus of elasticity; (**d**) inner cohension; (**e**) internal friction angle.

3.2.3. Thermal Test

As can be seen from Figure 9, the maximum and minimum of the thermal conductivity of 35 site soils is 1.092 and 0.548 W/m/K, and the average (marked with the red dashed line and the red arrow and text in Figure 9a) is 0.812 W/m/K; the maximum and minimum of the coefficient of thermal diffusion of 35 site soils is 1.190 and 0.357 m^2/s, and the average (marked with the red dashed line and the red arrow and text in Figure 9b) is 0.689 m^2/s; the maximum and minimum of the specific heat capacity of 35 site soils is 0.9813 and 0.4592 kJ/kg/K, and the average (marked with the red dashed line and the red arrow and text in Figure 9c) is 0.755 kJ/kg/K; the maximum and minimum of the coefficient of heat expansion of 35 site soils is 15.10×10^{-6} and 1.83×10^{-6} 1/K, and the average (marked with the red dashed line and the red arrow and text in Figure 9d) is 8.29×10^{-6} 1/K.

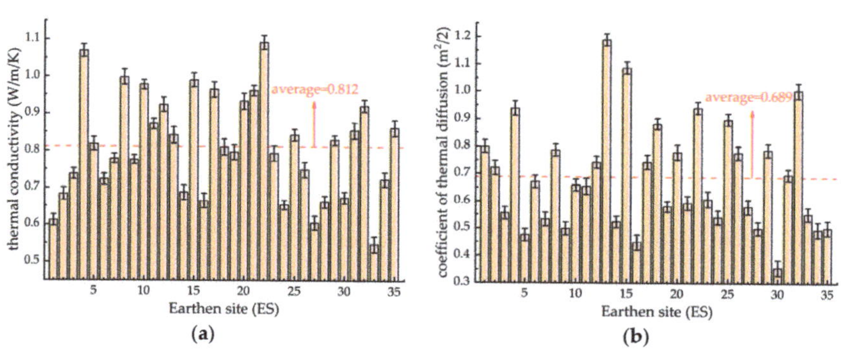

Figure 9. *Cont.*

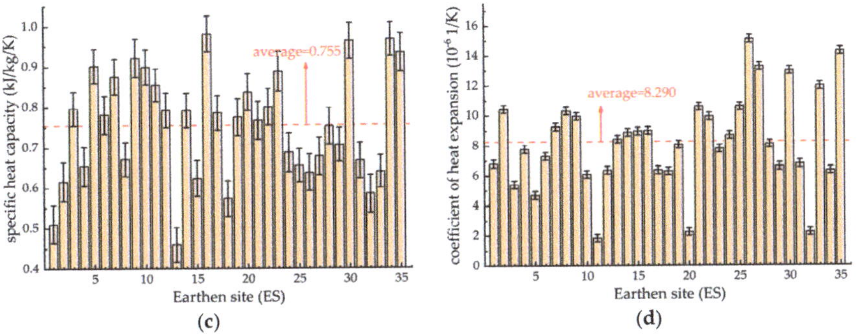

Figure 9. Thermal properties of 35 site soils. (**a**) Thermal conductivity; (**b**) coefficient of thermal diffusion; (**c**) specific heat capacity; (**d**) coefficient of heat expansion.

3.2.4. Water-Physical Test

As can be seen from Figure 1, the maximum and minimum of the permeability coefficient of 35 site soils is 69.7×10^{-6} and 1.7×10^{-6} cm/s, and the average (marked with the red dashed line and the red arrow and text in Figure 10a) is 26.671×10^{-6} cm/s; the maximum and minimum of the coefficient of thermal diffusion of 35 site soils is 1.67 and 0.001 g/min, and the average (marked with the red dashed line and the red arrow and text in Figure 10b) is 0.353 g/min.

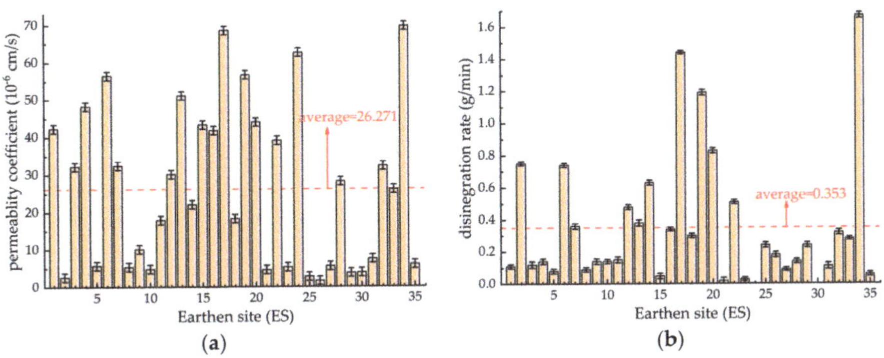

Figure 10. Water-physical properties of 35 site soils. (**a**) Permeability coefficient; (**b**) disintegration rate.

4. Analysis of Compatibility Assessment

There are three steps to accomplish the compatibility assessment of the slurries and site soils in NW China. The first step is to select indicators; the second step is to weigh indicators; the third step is to sort all of the samples. Careful consideration should be given to every step.

4.1. Select Indicators

Compatibility assessment needs to select assessment indicators before transferring to a multi-criteria decision-making model (MCDM). There are fifteen indicators of all properties in the previous results, but some indicators are closely related and some are not important. So, twelve indicators are selected after careful consideration of importance and independence: density, specific gravity, porosity, compression strength, modulus of elasticity, inner cohesion, internal friction angle, thermal conductivity, specific heat capacity, coefficient of heat expansion, permeability coefficient and disintegration rate. The following is the detail.

In the basic physical branch, density, specific gravity, and porosity are selected. Compared with other construction materials, the distinct character of soil is the three-phase composition. Basic physical indicators indicate the global property and the proportion of each phase; specially, the density indicates the global property, and the specific gravity and the porosity indicate the feature of solid phase and gas phase. Besides, other basic physical indicators can be expressed by the three indicators. Note that the water content is not selected since the water content of the site soil is closely related to the climate of the area, and when the slurry is applied in same area, it is possible that the slurry will reach the same value of water content as the soil.

In the mechanical branch, four indicators are selected, namely compressive strength, modulus of elasticity, inner cohesion, and internal friction angle. The compressive strength and modulus of elasticity are common and necessary parameters for mechanical calculations. According to the Mohr-Coulomb criterion, inner cohesion and internal friction angle are essential parameters to judge the state of stressed soil. Note that the tensile strength is strongly related with the compression strength in much of the literature, which shows the compression and tensile strength can substitute for each other in some degree.

In the thermal branch, thermal conductivity, specific heat capacity, and thermal expansion coefficient are selected. The thermal conductivity and specific heat capacity are important parameters in the calculation of heat. Considering the temperature change, the deformation is expressed by the coefficient of liner expansion. Note that the coefficient of thermal diffusion is not selected since it can be converted by thermal conductivity, specific heat capacity, and density in mathematical expression.

In the water-physical branch, permeability coefficient and disintegration rate are selected. The permeability coefficient shows the permeability in water saturation, and the disintegration rate shows the stability in water. In Northwest China, sudden rainfall in summer and snowfall in winter may lead to the rise of water levels, and further lead to a decline in strength and water movement. Two selected parameters are used to described a range of possibilities.

4.2. Weight Indicators

At present, the methods of weighting indicators can be divided into two main categories in general: subjective weights and objective weights. Analytic hierarchy process (AHP) is a common and basic method to calculate subjective weights; and criteria importance through intercriteria correlation (CRITIC) is a common method to calculate objective weights. Both methods are useful and efficient in some particular situations. So, two methods are adopted and then a combined weight is generated by subjective and objective weights.

4.2.1. AHP

The key of the AHP method is to construct the hierarchy model and ensure the importance of each index. The former is easy in the paper since every layer of the hierarchy model is clear after selecting indicators. The latter is decided by exports. The following is the detail.

Step 1: construct the hierarchy model.

In the model, compatibility assessment is the goal layer; there are four branches in the criterion layer, and there are 12 selected indicators in the index layer. Figure 11 shows the hierarchy model.

Step 2: ensure the weight decision matrix of each layer.

The weight decision matrix A can be given as follows:

$$A = (a_{ij})_{n \times n} \quad (1)$$

where a_{ij} is governed by the rules: $a_{ij} > 0$; $a_{ij} = 1/a_{ji}$ $(I \neq j)$; $a_{ij} = 1$ $(i = j = 1, 2, \ldots, n)$.

a_{ij} indicates a quantified judgment on a pair of elements A_i and A_j, which indicate sets of elements. The value of a_{ij} was determined from a nine-point scale, indicating preferences between options as equally, moderately, strongly, very strongly, and extremely preferred. These preferences can be expressed as pairwise weights of 1, 3, 5, 7, and 9, while 2, 4, 6, and 8 are intermediate values [31].

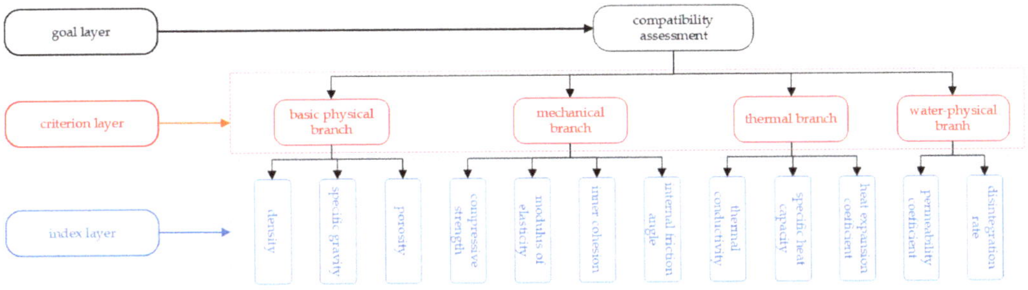

Figure 11. The hierarchy model of AHP.

Step 3: calculate the weight of each layer

The weight of each layer is gained from the eigenvector corresponding the maximum eigenvalue of the matrix A [32].

$$Aw = \lambda_{\max} w \quad (2)$$

where $w = (w_1, w_2, \ldots, w_n)$ is the eigenvector, A is the weight decision matrix, and λ_{max} is the largest eigenvalue of matrix A. Here, the authors use the eigenvector function in MATLAB to calculate w and λ_{max}.

Step 4: check the consistency

The aim of checking the consistency is to ensure that each pairwise comparison is consistent with others and gain an accurate result.

The consistency index (CI) was calculated by Equation (3):

$$CI = \frac{\lambda_{\max} - n}{n - 1} \quad (3)$$

The random index (RI) and its values for different scales of matrices are cited in the literature [33].

The judgmental consistency of matrix A can be obtained by calculating the consistency ratio (CR):

$$CR = \frac{CI}{RI} \quad (4)$$

When $CR \leq 0.1$, the matrix can be regarded as consistent. The final comparison matrix is shown in Appendix A.

Step 5 calculate the weight of each indicator.

The weight of each indicator is the multiplication of the weight of criteria layer and the weight of index layer, shown in Table 5.

Table 5. Subjective weight of each indicator.

Criterion Layer	Weight of Criterion Layer	Index Layer	Weight of Index Layer	Subjective Weight
Basic physic branch	0.0838	Density	0.1635	0.0137
		Specific gravity	0.2970	0.0249
		Porosity	0.5396	0.0452
Mechanical branch	0.5462	Compressive strength	0.0874	0.0477
		Modulus of elasticity	0.1428	0.0780
		Inner cohesion	0.3849	0.2102
		Internal friction angle	0.3849	0.2102

Table 5. *Cont.*

Criterion Layer	Weight of Criterion Layer	Index Layer	Weight of Index Layer	Subjective Weight
Thermal branch	0.1377	Thermal conductivity	0.5	0.0689
		Specific heat capacity	0.25	0.0344
		Coefficient of heat expansion	0.25	0.0344
Water-physical branch	0.2323	Permeability coefficient	0.25	0.0581
		Disintegration rate	0.75	0.1742

4.2.2. CRITIC

The key of the CRITIC method is the calculation of the amount of information [31]. The detailed steps of CRITIC are the followings.

Step 1: standardize the raw data.

Let i = 1~35 represent site soils, and j = 1~12 represent the evaluation indicators, and then standard the raw data, as Equation (5).

$$x_{ij} = \frac{r_{ij} - \min_j(r_{ij})}{\max_j(r_{ij}) - \min_j(r_{ij})} \tag{5}$$

Step 2: calculate the standard deviation and correlation coefficient.

Let δ_j represent the standard deviation of an indicator, and let cc_{jh} represent the correlation coefficient of two indicators.

Step 3: calculate the amount of information I_j.

$$I_j = \delta_j \sum_{h=1}^{12} (1 - cc_{jh}) \tag{6}$$

Step 4: compute the weight of each indicator.

$$W_j = \frac{I_j}{\sum_{j=1}^{12} I_j} \tag{7}$$

Finally, the combined weight is the average of the subjective and objective weights, and the result is shown in Table 6.

Table 6. Results of combined weight.

Indicators	Subjective Weight	Objective Weight	Combined Weight
Density	0.0137	0.0690	0.0414
Specific gravity	0.0249	0.0716	0.0482
Porosity	0.0452	0.0951	0.0702
Compressive strength	0.0477	0.0765	0.0621
Modulus of elasticity	0.0780	0.0855	0.0817
Inner cohesion	0.2102	0.0821	0.1461
Internal friction angle	0.2102	0.0732	0.1417
Thermal conductivity	0.0689	0.0848	0.0768
Specific heat capacity	0.0344	0.0806	0.0575
Coefficient of heat expansion	0.0344	0.0839	0.0591
Permeability coefficient	0.0581	0.1123	0.0852
Disintegration rate	0.1742	0.0856	0.1299

To facilitate subsequent matrix operations, the weight vector is converted to the diagonal matrix W, as follows:

$$W = diag(w_1, w_2, \cdots, w_{13}) \tag{8}$$

4.3. Sort All Samples

To sort all samples, there are many methods (i.e., technique for order preference by similarity to ideal solution (TOPSIS), vlse kriterijumska optimizacijal kompromisno resenje (VIOKR), preference ranking organization method for enrichment evaluations (PROMETHEE), west-worst method (BWM), decision making trail and evaluation laboratory (DEMATEL), elimination et choix traduisant la realite (ELECTRE), etc.) [34,35]. The most simple and basic method is TOPSIS, which is adopted in the paper.

The key in the application of TOPSIS is to determine what the ideal solution is in practice. According to the previous section, "absolute compatibility" is just wishful thinking, which means the absolute compatible material cannot be found and cannot represent the ideal solution. But the similarity is certainly an equivalent concept of the compatibility in some particular cases (i.e., the conservation material that is the most similar to the relic is sometimes considered to be the most compatible materials). This can be attempted in the assessment of compatibility between the slurry and earthen sites. The purposes of fissure grouting are: (1) to fill and seal fissures; and (2) to improve the continuity and stability of the earthen site. The achievement of purposes may be affected by the reason that properties of the slurry are obvious contrast with the soil, which leads to stress concentration and water-salt accumulation at the interface between slurry and soil. To avoid these, the slurry that is the most similar to the soil is considered as the most compatible material and the ideal solution. Certainly, it has to be realized that the aim is not to copy the site soil completely, but to use the possibility of obtaining appropriate properties by varying several parameters, and therefore optimize the slurry for its specific application [36]. Also, it is realized that the ideal solution should be different and determined again in other practices.

The following is the detailed steps of TOPSIS.

Step 1: build a standardized data matrix.

The raw data matrix R is composed of all values of 12 indicators of the slurry and 35 site soils. Note that the column represents the same indicator and the row represents the same site soil. In particular, the first row represents the slurry.

$$R = (r_{ij})_{36\times 12} \tag{9}$$

Then, calculate the absolute value of the difference in each indicator between site soils and the slurry:

$$b_{ij} = |r_{ij} - r_{1j}| \tag{10}$$

$$B = (b_{ij})_{36\times 13} \tag{11}$$

All data in matrix B need to be normalized to form the normalized matrix C:

$$c_{ij} = \frac{\max_j(b_{ij}) - b_{ij}}{\max_j(b_{ij}) - \min_j(b_{ij})} \tag{12}$$

Step 2: calculate the weighted normalized matrix D:

$$D = CW \tag{13}$$

Step 3: calculate the distances between the ideal solution and each sample.

As mentioned earlier, the slurry is considered the ideal solution. Specifically, the positive ideal solution is placed in the first row of the matrix D, and the negative ideal solution is the set of the value 0 in every column. So, the distances between the ideal solution and every sample can be expressed as shown:

$$\begin{aligned} L_i^+ &= \sqrt{\sum_{j=1}^{n} (d_{ij} - d_j^+)^2} = \sqrt{\sum_{j=1}^{n} (d_{ij} - d_{1j})^2} \\ L_i^- &= \sqrt{\sum_{j=1}^{n} (d_{ij} - d_j^-)^2} = \sqrt{\sum_{j=1}^{n} d_{ij}^2} \end{aligned} \tag{14}$$

Step 4: calculate the result of assessment.

The result of compatibility assessment is expressed as the relative distance, as shown (12):

$$E_i = \frac{L_i^-}{L_i^+ + L_i^-} \tag{15}$$

Finally, the global flow chart of compatibility assessment is shown in Figure 12.

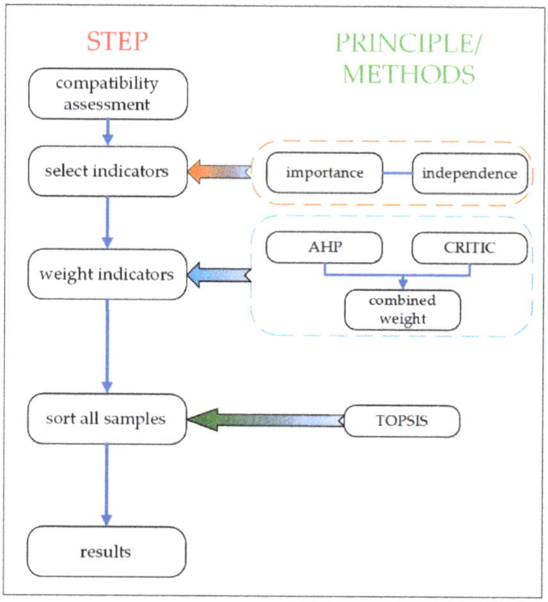

Figure 12. Flow chart of compatibility assessment.

4.4. Assessment Results

For the convenience of calculation, the computation is designed with MTABLAB, and the code is shown in Appendix B. Figure 13 shows the results of compatibility assessment. As can be seen from Figure 13a,b, (1) for one earthen site, the difference of the compatibility assessment among the three slurries is small since the properties of slurries are close; (2) for all earthen sites, the difference of the compatibility assessment of one slurry is very obvious; (3) most results of assessment range 0.4~0.8 (middle and high), few results range 0.2~0.4 (low) and 0.8~1 (very high), and no results range 0~0.2 (very low).

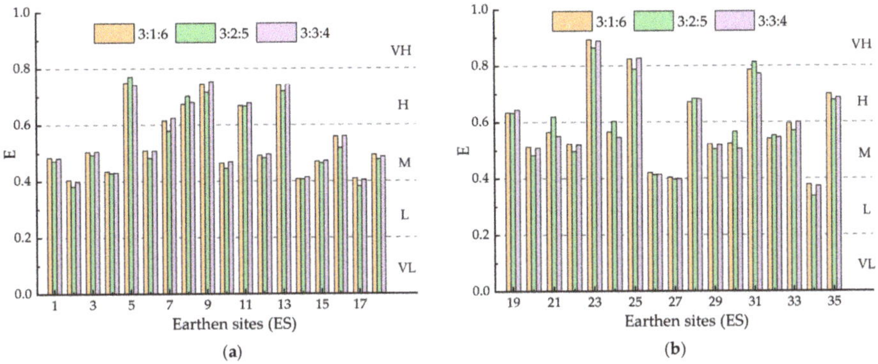

Figure 13. Results of compatibility assessment. (**a**) ES1~18; (**b**) ES19~35.

5. Discussion

The importance of compatible slurries in the conservation practice for earthen sites is obvious. So, the essential and necessary work which must be carried out is to get the properties of the slurries and to assess their compatibility. Firstly, the properties should be comprehensive. It is understood that the basic physical and mechanical properties must be considered and tested for any kind of engineering. The additional properties should be considered and tested considering the actual application environment involved and the aim of long-term service. In NW China, strong sunshine and sudden rainfall are obvious climatic characteristic, and hence the slurries and earthen sites are easily exposed to strong sunlight and local excess water. Considering the above, thermal and water-physical properties should be test when the slurries may be applied in NW China. Certainly, the same tests should be taken for the site soils. The results show the properties of site soils are obviously different, and furthermore that one slurry could not be adjusted to all earthen sites. Then, the compatibility assessment needs to be considered. MCDM seems to be a universal model to assess the compatibility. To process MCDM, some basic principles and simple steps are clear, but the details should be careful considered in a particular conservation practice. A basic physical branch, mechanical branch, thermal branch, and water-physical branch are considered, and twelve indicators are selected in the paper. Two methods are adopted to gain a combined weight of each indicator, and then the TOPSIS method is used to sort all samples. The reasonable selections and weighting are the basic premise of get correct results. When it is necessary to evaluate the compatibility of another conservation material in different regions, the steps in the paper are worthy of reference. Finally, it should be realized that the aim to get the full-scale properties and to assess the compatibility is to ensure that the compatible materials could perform well in the appropriate environment.

6. Conclusions

Slurries for fissure restoration in earthen sites must be compatible with site soils. The content of the compatibility includes more than one branch. In northwest China, the slurry has to meet more requirements for the particularly arid environment. Besides the basic physical and mechanical properties of the slurry, thermal and water-physical properties should be considered, and the corresponding properties of the site soils have to be tested to compare them more easily. Meanwhile, the compatibility assessment between the slurry and the site soil is carried out. Also, we have realized that comparisons with other similar types of slurry are necessary and need to be researched in the future. The main conclusions are as follows:

(1) The slurries with quicklime for fissure restoration in NW China are tested in multi-properties, and the properties of the slurries with different mix proportions are close; the slurry with the mix proportion 3:2:5 generally is considered as the better slurry.

(2) Thirty-five typical earthen sites in NW China are selected and the same tests are taken; the range of each properties for all site soils is clear and the difference are obvious.

(3) Compatibility assessment can be considered as MCDM; to proceed with the work, three necessary and basic steps are involved: to select indicators of compatibility assessment, to weigh each indicator, and to sort all of the samples; these steps can be easily adopted in the similar practice.

(4) The principle of indicators selection is the importance and the dependence; the combined weight is the average of subjective and objective weights that are decided by AHP and CRITIC; the TOPSIS method is used to sort all samples.

(5) Most results of compatibility assessment are middle (M) and high (H), few results are very high (VH) and low (VL), and no results are very low (VL). The three slurries can be considered being compatible with most earthen sites.

Author Contributions: Conceptualization, X.Y. and K.C; methodology, X.Y. and X.H.; software, X.Y.; investigation, G.Z.; data curation, Q.Q.; writing—original draft preparation, X.Y.; writing—review and editing, K.C.; project administration, X.Y. and K.C; funding acquisition, K.C. All authors have read and agreed to the published version of the manuscript.

Funding: This research was funded by the National Nature Science Foundation of China (52068050).

Institutional Review Board Statement: Not applicable.

Informed Consent Statement: Not applicable.

Data Availability Statement: Not applicable.

Conflicts of Interest: The authors declare no conflict of interest.

Appendix A

Table A1. The following is the comparison matrix of criterion/index layer.

Criterion/Index Layer	Comparison Matrix				Weight
Criterion Layer					
	(1)	(2)	(3)	(4)	
(1) basic physic branch	1	1/5	1/2	1/3	0.0838
(2) mechanical branch	5	1	4	3	0.5462
(3) thermal branch	2	1/4	1	1/2	0.1377
(4) water-physical branch	3	1/3	2	1	0.2323
λmax = 3.7321, CR = 0.0189 < 0.1					
Index Layer					
basic physic branch	(1)	(2)	(3)		
(1) density	1	1/2	1/3		0.1635
(2) specific gravity	2	1	1/2		0.2970
(3) porosity	3	2	1		0.5396
λmax = 3.0092, CR = 0.0079 < 0.1					
mechanical branch	(1)	(2)	(3)	(4)	
(1) compressive strength	1	1/2	1/4	1/4	0.0874
(2) modulus of elasticity	2	1	1/3	1/3	0.1428
(3) inner cohesion	4	3	1	1	0.3849
(4) internal friction angle	4	3	1	1	0.3849
λmax = 4.0206, CR = 0.0076 < 0.1					
thermal branch	(1)	(2)	(3)		
(1) thermal conductivity	1	2	2		0.5
(2) specific heat capacity	1/2	1	1		0.25
(3) heat expansion coefficient	1/2	1	1		0.25
λmax = 3, CR = 0 < 0.1					
water-physical branch	(1)	(2)			
(1) permeability coefficient	1	1/3			0.25
(2) disintegration rate	3	1			0.75
λmax = 2, CR = 0 < 0.1					

Appendix B

The following is the MATLAB code of the realization of the compatibility assessment.

```
clc;
clear all;
SO=xlsread('soil.xlsx');
SL=xlsread('slurry.xlsx');
for j=1:12;
    maxc1(j)=max(SO(:,j));
    minc1(j)=min(SO(:,j));
    CRSTD(:,j)=(maxc1(j)-SO(:,j))/((maxc1(j)-minc1(j)));
end
s=std(CRSTD);
cor=corrcoef(CRSTD);
qi=s.*sum(1-cor);
wo=qi/sum(qi);
wa=[0.0137 0.0249 0.0452 0.0477 0.0780 0.2102 0.2102 0.0689 0.0344 0.0344 0.0581 0.1742];
W=diag((wa+wo)/2);
for jj=1:3;
    for i=1:35;
        B(i,:)=abs(SO(i,:)-SL(jj,:));
    end
    for j=1:12;
        maxc2(j)=max(B(:,j));
        minc2(j)=min(B(:,j));
        C(:,j)=(maxc2(j)-B(:,j))/((maxc2(j)-minc2(j)));
    end
    D=C*W;
    for j=1:12;
        maxr2(j)=max(D(:,j));
        minr2(j)=min(D(:,j));
        LP(:,j)=(D(:,j)-maxr2(j)).^2;
        LN(:,j)=D(:,j).^2;
    end
    DLP=sqrt(sum(LP,2));
    DLN=sqrt(sum(LN,2));
    E(jj,:)=DLN./(DLP+DLN);
end
plot(E.');
```

References

1. The Getty Conservation Institute, Pontificia Universidad Católica del Perú. *Interdisciplinary Experts Meeting on Grouting Repairs for Large-Scale Structural Cracks in Historic Earthen Buildings in Seismic Area*; The Getty Conservation Institute: Los Angeles, CA, USA, 2007.
2. Pearson, G.T. *Conservation of Clay and Chalk Buildings*; Donhead Publishing Ltd.: Shaftsbury, VT, USA, 1997.
3. Keefe, L. *Earth Building: Methods and Materials, Repair and Conservation*; Taylor & Francis: London, UK, 2005.
4. Zhang, J.; Chen, W.; Li, Z.; Wang, X.; Guo, Q.; Wang, N. Study on workability and durability of calcined ginger nuts-based grouts used in anchoring conservation of earthen sites. *J. Cult. Herit.* **2015**, *16*, 831–837. [CrossRef]
5. González-Sánchez, J.F.; Fernández, J.M.; Navarro-Blasco, Í.; Alvarez, J.I. Improving lime-based rendering mortars with admixtures. *Constr. Build. Mater.* **2021**, *271*, 121887. [CrossRef]
6. Lv, J.; Zhou, T.; Du, Q.; Wu, H. Experimental investigation on properties of gypsum-quicklime-soil grout material in the reparation of earthen site cracks. *Constr. Build. Mater.* **2017**, *157*, 253–262. [CrossRef]
7. Cui, K.; Yu, X.; Huang, J.; Shao, H.; Gu, X. Early solidification process of the slurry with quicklime for fissure restoration of earthen sites. *J. Build. Eng.* **2022**, *53*, 104530. [CrossRef]

8. Cui, K.; Feng, F.; Chen, W.-W.; Wang, D.-H.; Wang, X.-H. Slurry and Technology Optimization for Grouting Fissures in Earthen Sites with Quicklime. *Adv. Mater. Sci. Eng.* **2019**, *2019*, 1–11. [CrossRef]
9. Rodrigues, J.D.; Grossi, A. Indicators and ratings for the compatibility assessment of conservation actions. *J. Cult. Herit.* **2007**, *8*, 32–43. [CrossRef]
10. Barone, G.; Branca, C.; Gresta, S.; Imposa, S.; Leone, A.; Majolino, D. Geoarcheometric and geophysical methodologies applied to the study of cultural heritage: "St. Agata la Vetere" in Catania (Sicily, Italy). *J Cult Herit.* **2004**, *5*, 263–271. [CrossRef]
11. Feng, Z.; Luo, X.; Wang, J.; Cao, S.-J. Energy-efficient preservation environment control for enclosed exhibition hall of earthen relics. *Energy Build.* **2021**, *256*, 111713. [CrossRef]
12. Zhang, C.; Zhang, B.; Cui, B. High hydrophobic preservation materials can cause damage to tabia relics. *Prog. Org. Coatings* **2020**, *145*, 105683. [CrossRef]
13. Zhang, T.; Xu, H.; Wang, C. Self-adaptability and topological deformation of Ganlan architectural heritage: Conservation and regeneration of Lianghekou Tujia village in Western Hubei, China. *Front. Arch. Res.* 2022; *in press*. [CrossRef]
14. Fan, W.; Chen, W.; Zhang, Q.; Wu, G. Feasibility of protecting earthen sites with sticky rice and lime composite. *Constr. Build. Mater.* **2022**, *346*, 128449. [CrossRef]
15. Rodrigues, J.D. INCOMARECH Project: Conclusions and tentative recommendations. *PACT* **1999**, 56.
16. Cui, K.; Feng, F.; Chen, W.; Wang, X.; Cheng, F. Study on the mechanical compatibility of fissure grouting slurry with quick lime and grouting technology optimization in earthen sites. *Rock Soil Mech.* **2019**, *40*, 1–10. (In Chinese)
17. Loureiro, A.M.S.; Paz, S.P.A.; Veiga, M.D.R.; Angélica, R.S. Assessment of compatibility between historic mortars and lime-METAKAOLIN restoration mortars made from amazon industrial waste. *Appl. Clay Sci.* **2020**, *198*, 105843. [CrossRef]
18. Sáez-Pérez, M.P.; Durán-Suárez, J.A.; Verdú-Vázquez, A.; Gil-López, T. Characterization and chromatic evaluation of gypsum-based pastes for construction and heritage restoration. *Constr. Build. Mater.* **2021**, *307*, 124981. [CrossRef]
19. Lubelli, B.; Nijland, T.G.; van Hees, R.P.J. Characterization and compatibility assessment of commercial stone repair mortars. *J. Cult. Herit.* **2021**, *49*, 174–182. [CrossRef]
20. Sala, E.; Zanotti, C.; Passoni, C.; Marini, A. Lightweight natural lime composites for rehabilitation of Historical Heritage. *Constr. Build. Mater.* **2016**, *125*, 81–93. [CrossRef]
21. Mazzuca, C.; Severini, L.; Domenici, F.; Toumia, Y.; Mazzotta, F.; Micheli, L.; Titubante, M.; Di Napoli, B.; Paradossi, G.; Palleschi, A. Polyvinyl alcohol based hydrogels as new tunable materials for application in the cultural heritage field. *Colloids Surfaces B: Biointerfaces* **2020**, *188*, 110777. [CrossRef]
22. Bal, D.K.; Chandan, M.R.; Taneja, R.; Tiwari, R.R.; Saboor, S.; Mishra, D.; Ghosh, A.; Shaik, A.H. Biocompatible polymer-capped oxidation-resistant copper nanoparticles for nanofluid and hydrogel applications. *Eur. Phys. J. Plus* **2021**, *137*, 1–16. [CrossRef]
23. Du, Y.; Chen, W.; Cui, K.; Zhang, K. Study on Damage Assessment of Earthen Sites of the Ming Great Wall in Qinghai Province Based on Fuzzy-AHP and AHP-TOPSIS. *Int. J. Arch. Heritage* **2019**, *14*, 903–916. [CrossRef]
24. Kim, C.J.; Yoo, W.S.; Lee, U.K.; Song, K.J.; Kang, K.I.; Cho, H. An experience curve-based decision support model for prioritizing restoration needs of cultural heritage. *J. Cult. Herit.* **2010**, *11*, 430–437. [CrossRef]
25. Kutut, V.; Zavadskas, E.K.; Lazauskas, M. Assessment of priority alternatives for preservation of historic buildings using model based on ARAS and AHP methods. *Arch. Civ. Mech. Eng.* **2014**, *14*, 287–294. [CrossRef]
26. Li, J.; Chen, Y.; Yao, X.; Chen, A. Risk Management Priority Assessment of heritage sites in China Based on Entropy Weight and TOPSIS. *J. Cult. Herit.* **2021**, *49*, 10–18. [CrossRef]
27. Wang, Y.M.; Yang, Z.C.; Chen, W.W.; Han, W.F. Strength characteristics and mechanism of loess solidified with new polymer material SH. *Chin. J. Rock Mech. Eng.* **2005**, *24*, 2554–2559.
28. He, F.G.; Lv, R.; Su, H.Z.; Zhou, J.; Zhang, J.K.; Wang, N. Durability test and reinforced mechanism on adding SH materials into soil of archaeological sites. *Rock Soil Mech.* **2019**, *40*, 297–307. (In Chinese)
29. Guan, X.P. Study on the Correlation between fractal and engineering of rammed earth of the Ming Great Wall on the foot of Qilian Mountains. Master's Thesis, Lanzhou University of Technology, Lanzhou, China, 2017.
30. Du, Y.M. Military defense system and vulnerability assessment of earthen sites of the Ming Great Wall in Qinghai Province. Ph.D. Thesis, Lanzhou University, Lanzhou, China, 2019.
31. Saaty, T.L. Rank from comparisons and from ratings in the analytic hierarchy/network processes. *Eur. J. Oper. Res.* **2006**, *168*, 557–570. [CrossRef]
32. Saaty, T.L. *The Analytic Hierarchy Process: Planning, Priority Setting, Resources Allocation*; McGraw: New York, NY, USA, 1980. [CrossRef]
33. Saaty, T.L. The analytic hierarchy process: A 1993 overview. *Cent. Eur. J. Oper. Res. Econ.* **1993**, *2*, 119–137.
34. Alao, M.; Ayodele, T.; Ogunjuyigbe, A.; Popoola, O. Multi-criteria decision based waste to energy technology selection using entropy-weighted TOPSIS technique: The case study of Lagos, Nigeria. *Energy* **2020**, *201*, 117675. [CrossRef]
35. Asante, D.; Ampah, J.D.; Afrane, S.; Adjei-Darko, P.; Asante, B.; Fosu, E.; Dankwah, D.A.; Amoh, P.O. Prioritizing strategies to eliminate barriers to renewable energy adoption and development in Ghana: A CRITIC-fuzzy TOPSIS approach. *Renew. Energy* **2022**, *195*, 47–65. [CrossRef]
36. Zacharopoulou, G. The renascence of lime based mortar technology: An appraisal of a bibliographic study. *Compat. Mater. Prot. Eur. Cult. Herit.* **1998**, *55*, 89–114.

coatings

MDPI

Article

The Influence of Natural Aging Exerting on the Stability of Some Proteinaceous Binding Media Commonly Used in Painted Artworks

Zhenzhen Ma [1,2,3], Lu Yang [4,5,6], Liqin Wang [4,5,6], Václav Pitthard [7], Tatjana Bayerova [8], Gabriela Krist [8] and Xichen Zhao [1,2,3,*]

1 Department of Conservation Research, Shaanxi Academy of Archaeology, Xi'an 710054, China
2 Key Scientific Research Base of On-Site Conservation, State Administration for Cultural Heritage, Xi'an 710054, China
3 Shaanxi Key Laboratory of Archaeological Conservation, Xi'an 710054, China
4 China-Central Asia "the Belt and Road" Joint Laboratory on Human and Environment Research, Xi'an 710054, China
5 Key Laboratory of Cultural Heritage Research and Conservation, Northwest University, Ministry of Education, Xi'an 710054, China
6 School of Cultural Heritage, Northwest University, Xi'an 710054, China
7 Department of Conservation Science, Kunsthistorisches Museum Wien, 1010 Vienna, Austria
8 Department of Conservation and Restoration, University of Applied Arts Vienna, 1010 Vienna, Austria
* Correspondence: zxc0324@163.com

Abstract: Natural aging plays a key role in the degradation of proteinaceous binders which are important dispersers and stabilizers of painting layers. Knowledge about the natural aging influence on the stability of binders is important for exploring the deterioration mechanisms of painted artworks. Pig glue, whole egg, egg white, egg yolk, casein, and mixtures with ultramarine were aged for 10 years in natural conditions. GC-MS and FTIR were applied to explore the changes in the binders at a molecular level. Our experiment revealed that the less stable Met (Methionine), Lys (Lysine), Ile (Isoleucine), Ser (Serine), Asp (Aspartic acid), Glu (Glutamic acid), Hyp (Hydroxyproline), especially aromatic Phe (Phenylalanine), and Tyr (Tyrosine) were damaged, thus the contents of the stable Ala (Alanine), Gly (Glycine), Val (Valine), Leu (Leucine), and Pro (Proline) increased. The broadening of Amide A and the declining amount of α-helix, along with the increasing contents of β-sheet and random coils, all showed that the binders had transformed into disordered states. What is more, we found that pig glue had better natural aging resistance, ultramarine could speed up the aging process and lipids in egg were more easily degraded. The mechanisms of the changes of primary structures and secondary structures are also discussed in the paper.

Keywords: painted artworks; binding medium; natural aging; stability; GC-MS; FTIR

1. Introduction

Proteinaceous binding media play an important role in scattering and consolidating pigments, which ultimately affects the permanence of paint [1]. As typical organic biopolymers, binders are degraded by light and heat, which could cause irreversible photooxidation, photochemistry, and decomposition reactions, resulting in diseases such as efflorescence and detachment in painted artworks. Despite extensive research into proteinaceous binders under short-term artificial aging [2–4], there is a substantial lack of knowledge regarding the long-term natural aging influence on proteinaceous binders. In this paper, five common proteinaceous binders of painted artwork (pig glue, whole egg, egg white, egg yolk, and casein), as well as mixtures with ultramarine, were aged for 10 years in natural conditions. GC-MS and FTIR were applied to study the changes in primary structures (amino acids) and secondary structures (α-helix, parallel β-sheet, anti-parallel β-sheet, random coils, and β-turns) of the proteins at a molecular level in an attempt to explore their stability characteristics for resisting natural aging. This research is of great significance to the study of the degradation mechanism of painted artworks. Furthermore, it will help with decisions regarding proper conservation materials following the artwork conservation principle, "original materials, original technique".

2. Materials and Methods

2.1. Reagents and Chemicals

The reagents and chemicals used were pig glue (German Kremer, 63000, Aichstetten, Germany); egg (egg white and egg yolk were acquired by separating whole egg) from the market; casein (German Kremer, 63200, Aichstetten, Germany); ultramarine (German Kremer, 45040, Aichstetten, Germany); KBr. AAS 18 mixed standard amino acid solution (Sigma-Aldrich, St. Louis, MO, USA); L-Norleucine (Sigma-Aldrich, St. Louis, MO, USA); MTBSTFA (Sigma-Aldrich, St. Louis, MO, USA); pyridine (Guangfu Chemical Research Institute, Tianjin, China); ammonium hydroxide (Guangfu Chemical Research Institute, Tianjin, China); HCl (Xilong Chemical Company, Sichuan, China); He (99.999%, Messer Company, Sichuan, China); and ultrapure water (Master-S Hitech Science, Shanghai, China) [5].

2.2. Apparatus

GC-MS analysis was performed with a 7890A-5975C GC-MS (Agilent,, USA); ABJ220-4NM balance (Kern, Germany); an MS 3 DS25 vortex oscillator (IKS Klingelnberg GmbH); a KQ-50E ultrasonic cleaner (Kunshan Ultrasonic Instrument, Suzhou, China); an Anke LXJ-IIB centrifuge Anting Scientific Instrument Factory, Shanghai, China); QYN100-2 nitrogen purging apparatus (Qiaoyue Electrical Company, Shanghai, China); an MDS-8G microwave hydrolyzer (Xinyi Microwave Chemical Technology, Shanghai, China); a 10 μL injector (Agilent, Palo Alto, CA, USA); and an LS120A balance (Precisa, Zurich, Switzerland). FTIR analysis was performed with a Thermo Scientific Nicolet iN10 MX FTIR (Nicolet, Waltham, MA, USA), and a tablet press (Specac, Kent, UK), while a SOL 2 Xenon artificial aging box (Dr K. Hönle, Frickingen, Germany) was the aging apparatus.

2.3. Samples and Aging

Two sets of 3% (the proper concentration to mix pigments and paint with) pig glue, whole egg, egg white, egg yolk, and casein were deposited onto glass slides to form dry films. A third set of each binder with ultramarine was also made in the same way (simulating the painted layer composed of pigment and binders). The first set of the five pure binders was used at fresh standards, while the other two sets were aged under the SOL 2 Xenon artificial aging box, which had similar to daylight energy with a spectrum range of 300~800 nm. Illuminance was 120 lux, irradiation intensity was 910 $W \cdot m^{-2}$, lamp temperature was 50 °C, and aging time was 900 h (approximately 2 years' natural aging). After artificial aging, the aged samples were placed in the daily environment in the Kunsthistorisches Museum. Illuminance was 30~50 lux, irradiation intensity was 230~380 $W \cdot m^{-2}$, temperature was 16~32 °C, and aging time was 10 years.

2.4. Methods

2.4.1. GC-MS

EI positive mode (70 eV) was chosen in the MS spectrometer and the MS ion source temperature was 230 °C. The MS transfer line temperature was set at 280 °C and the MS quadrupole temperature at 150 °C. A HP-5MS fused silica capillary column (30 m × 0.25 mm × 0.25 μm) was applied for the separation, and the carrier gas flowed at 1.5 mL/min. The temperature oven program was performed from 100 °C (2 min) to 280 °C at 6 °C/min in spitless mode. The pretreatments of the samples made use of previously published working methodology and the five main steps (ammonia extraction of the protein, clean-up of the upper extraction with the C4 sorbent pipette tip, hydrolysis, derivatization, and injection) followed those laid out in [6] by our research group.

2.4.2. FTIR

The extraction of the fresh, aged–pure and aged–mixed binders was performed with 2 mL 2.5 mol/L ammonia for 1.5 h. The supernatant was then acquired after being centrifuged at 5000 rpm for 5 min, twice, which was then dried on KBr pellets for use. The measuring resolution was 4 cm^{-1} in a range of 675~4000 cm^{-1}, 64 times. The atmospheric interference was firstly deducted with Ominic 8.2 for baseline correction, then deconvolution of the Amide I band was performed (band enhancement factor: 3, FWHM: 42.2 cm^{-1}). By using a seven-point Savitzky-Golay function, the second order derivatives of the spectrogram could be obtained, after which the peaks of the five secondary structures (α-helix, parallel β-sheet, anti-parallel β-sheet, random coils, and β-turns) were identified. Finally, the contents of each secondary structure were calculated with the Gaussian curve fitting method.

3. Results and Discussion

3.1. GC-MS Results

3.1.1. Individual Amino Acids

The chromatograms of the five fresh, natural-aged pure and mixed proteinaceous binders are shown in Figure 1. Met, Lys, and Tyr all disappeared after aging, in accordance with the conclusions of Prof. Colombini [2] and Prof. Shuya Wei [4].

The average contents of the 11 amino acids are illustrated in Table 1. It could be observed that the contents of Ala, Gly, Val, Leu, and Pro showed climbing trends with a rough average increase of 2% and a maximum of 2.94% in Pro. The contents of Ile, Ser, Phe, Asp, Glu, and Hyp all decreased with the most obvious decline of 3.88% in Phe. What is more, we observed the most changes in amino acid content in egg yolk (−5.91%~4.62%), followed by egg white (−4.49%~3.59%), whole egg (−4.41%~3.35%), and casein (−4.31%~2.93%), and the fewest in pig glue (−1.55%~1.85%). What should also be stressed is that the contents of most amino acids changed more obviously in the mixed binders than in the pure ones.

Table 1. The average contents of the 11 amino acids in the five fresh and aged binders.

Samples	Ala	Gly	Val	Leu	Ile	Ser	Pro	Phe	Asp	Glu	Hyp
FPG	11.49	29.82	4.04	4.68	2.01	2.56	11.99	2.45	7.79	12.01	11.17
APG	12.35	30.62	5.71	5.87	1.38	1.40	13.12	1.10	7.14	11.12	10.18
APGU	12.75	31.02	5.47	6.09	1.03	1.11	13.84	0.90	6.94	11.00	9.86
FWE	9.68	5.73	6.31	12.91	4.39	12.16	5.79	7.99	16.52	18.52	0.00
AWE	12.06	8.07	8.71	14.99	2.96	9.11	8.92	3.77	14.92	16.49	0.00
AWEU	11.94	8.47	9.10	15.47	3.01	8.69	9.14	3.58	14.63	15.96	0.00
FEW	9.18	5.35	6.92	13.62	5.29	10.21	5.18	9.12	17.09	18.05	0.00
AEW	10.44	6.89	8.72	16.26	4.20	8.17	8.10	5.64	15.26	16.31	0.00
AEWU	10.60	7.60	9.40	15.85	3.93	7.48	8.77	4.63	15.66	16.08	0.00
FEY	10.69	6.12	5.98	11.41	4.55	13.58	6.99	6.93	15.15	18.62	0.00
AEY	12.70	10.04	9.25	14.72	3.35	10.10	10.60	1.25	12.27	15.72	0.00
AEYU	13.36	9.87	9.57	15.08	2.95	9.83	11.60	1.02	11.62	15.10	0.00
FC	5.20	3.12	8.71	13.03	7.23	4.35	15.58	7.15	10.12	25.53	0.00
AC	7.43	4.47	11.00	14.97	6.09	2.32	18.05	3.73	8.57	23.38	0.00
ACU	7.15	4.82	11.63	15.17	6.65	2.10	18.30	2.83	8.47	22.87	0.00

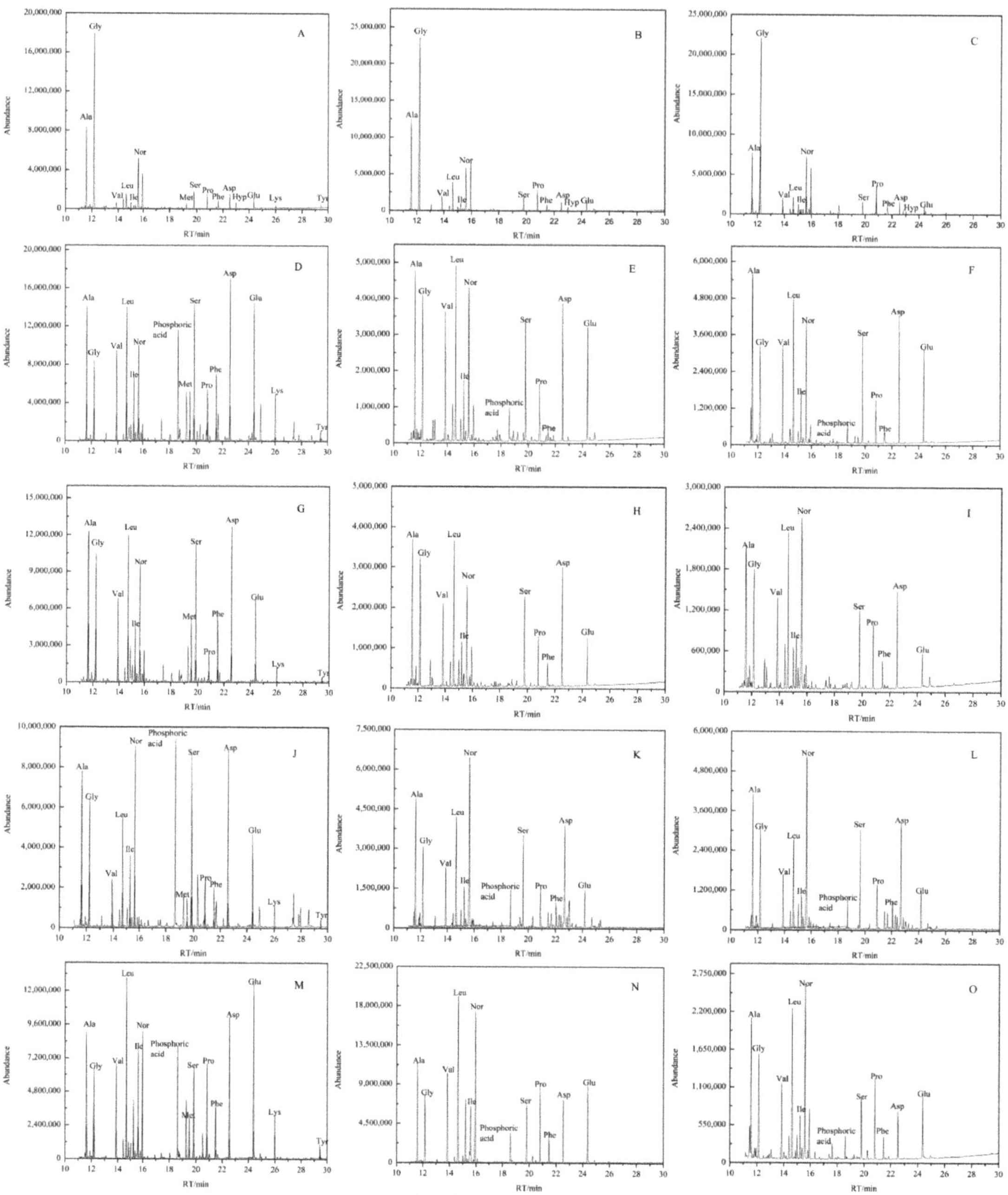

Figure 1. Chromatograms of the five fresh, natural-aged binders ((**A**) fresh pig glu-FPG, (**B**) aged pig glue-APG, (**C**) aged pig glue with ultramarine-APGU,(**D**): fresh whole egg-FWE, (**E**): aged whole egg-AWE, (**F**): aged whole egg with ultramarine-AWEU, (**G**): fresh egg white-FEW, (**H**): aged egg white-AEW, (**I**): aged egg white with ultramarine-AEWU, (**J**): fresh egg yolk-FEY, (**K**): aged egg yolk-AEY, (**L**): aged egg yolk with ultramarine-AEYU, (**M**): fresh casein-FC, (**N**): aged casein-AC, (**O**): aged casein with ultramarine-ACU).

3.1.2. Different Categories of Amino Acids

Normally, amino acids can be divided into seven groups according to their characteristics (hydrophilic amino acids: Asp, Glu, Ser, Hyp; electriferous hydrophilic amino acids: Asp, Glu; electrically neutral hydrophilic amino acids: Ser, Hyp; hydrophobic amino acids: Ala, Gly, Val, Leu, Ile, Pro, Phe; aliphatic amino acids: Ala, Gly, Val, Leu, Ile, Ser, Asp, Glu; aromatic amino acids: Phe; imino amino acids: Pro, Hyp) [7,8]. Their total content fluctuations were calculated and then shown in Figure 2. These calculations were in accord with the values reported in Table 1. The contents of hydrophilic, electriferous hydrophilic, electrically neutral hydrophilic amino acids, and aromatic amino acids were lowered with average deductions of 6.70%, 3.93%, 2.77%, and 3.88%, respectively. Whereas the contents of the hydrophobic, aliphatic (the increase of Ala, Gly, Val, Leu > the decrease of Ile, Ser, Asp, Glu), and imino amino acids (the increase of Pro > the decrease of Hyp) rose with average amounts of 6.70%, 1.17%, and 2.71%, respectively. Similar to the individual amino acids' changing trends, the variation range in egg yolk was the biggest (±10.80%), with those of whole egg (±7.91%), casein (±6.56%), and egg white (±6.13%) being smaller, and pig glue the smallest (±4.63%).

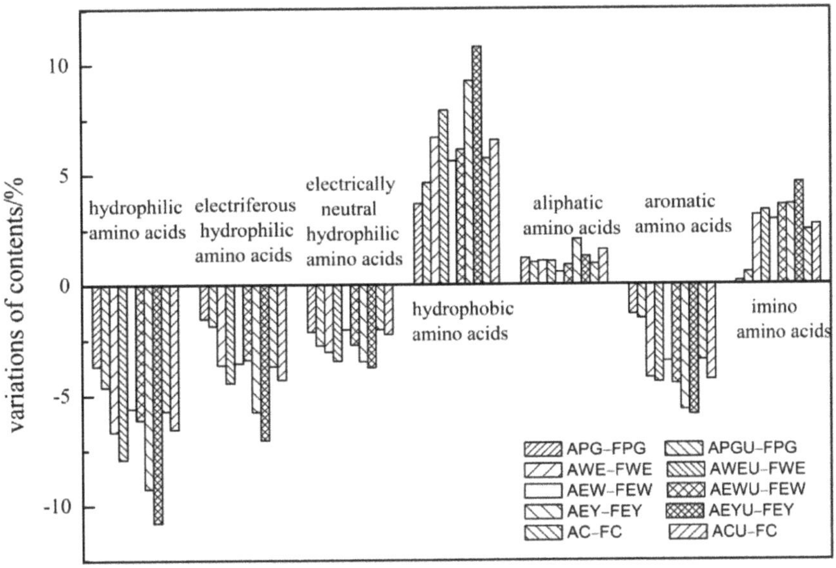

Figure 2. The content fluctuations of the seven groups of amino acids in the five fresh and aged binders.

3.1.3. Mechanisms of the Changes in Amino Acids

Light Aging

The chemical bonds in proteins would break down when the light energy exceeded the intermolecular hydrogen bond energy and the covalent bond energy between atoms, which would result in the denaturation and degradation of proteins [9]. Table 2 shows that the energy of the light wave ranging from 280 nm to 438 nm could destroy chemical bonds such as C–N, C–S, C–C, C–O, N–H, and C–H in the main and side chains of proteins and P–O in phosphoric acids [10]. Photolytic and photooxidation reactions would thus be caused [11] and the long chains would break into short chains, resulting in the decrease of their molecular weights.

Table 2. The energy of the different wavelengths and the chemical bonds in proteins.

Wavelength /nm	Energy /KJ·mol^{-1}	Chemical Bonds	Bond Energy /KJ·mol^{-1}
438	273	C–S	272
392	305	C–N	305
356	336	P–O	335
346	346	C–C	346
334	358	C–O	358
308	388	H–N	386
290	412	H–C	411

Moreover, free radicals (H, O, C, N, OH, CH_3) of high activity were produced during the above reactions and they could accelerate the deterioration process. For example, Davies found that the active oxygen radical would act with all side chains in amino acids and its concentration was positively correlated with the degree of destruction [12]. Among the 11 amino acids, the benzene ring in aromatic Phe and Tyr and the element S in Met enabled them to have high reactivity towards the free radicals [13]. What is more, it was easier for Phe and Tyr to absorb ultraviolet rays, perform dehydrogenation reactions, form color rendering conjugated systems, and result in obvious photooxidation reactions, due to their benzene rings [14]. Consequently, the content of Phe decreased markedly, and Tyr almost disappeared. In addition, yellow products with carbonyl structure were formed. Baltova and Vassileva also believed that the etiolation degree kept a linear correlation with the content of carbonyl structure [15]. It is important that related research has proven that the degradation of Phe and Tyr were the key factors resulting in light aging [16], inducing diseases such as yellowing and brittleness in artworks [17]. In addition, Phe could not absorb too many photons and thus degraded less [18], but the photons absorbed by Tyr could be transported to Phe and other amino acids, promoting even more deterioration actions and obvious light aging [19].

Thermal Aging

Long-term thermal aging could destroy the structure of protein and cause diseases such as denaturation, shrinkage, brittleness, and yellowing [20]. Large amounts of research have illustrated that the thermal stabilities of the common amino acids are (Val, Leu) > Ile > Tyr > Lys > Met > Ser > (Asp, Glu) [21]. Consequently, the content of Asp, Glu, Ser, Met, Lys, Tyr, and Ile decreased while that of Val and Leu increased. The specifics are as follows: (1) The electriferous amino acids were more quickly aged than the aliphatic amino acids [22]. The stabilities of the proteins were correspondingly weakened with the decline of Asp, Glu, and Lys [23]. (2) The –OH and phenolic hydroxyl group in Ser and Tyr contributed to oxidation, which would aggravate deterioration [24]. The degradation of serine phosphate was also an important factor in the decline of Ser (Ser existed in the form of serine phosphate in egg and casein). (3) Hyp was much easier to be degraded due to oxidation caused by active –OH. (4) The experiment showed that there was a relationship between Met and the thermal stability of proteins. The sulfydryl in Met helped to form a disulfide bond which could maintain spatial conformation and contribute to thermal stability; the higher the content of S-containing amino acids, the better the thermal stability [25]. However, due to the aging process, the disulfide bonds would break down which was not helpful in stabilizing the proteins.

Based on the above light and thermal aging discussions, the contents of Ile, Ser, Asp, Glu, Lys, Hyp, Met, and especially aromatic Phe and Tyr, showed decreasing trends, while those of Ala, Gly, Val, Leu, and Pro climbed. The aliphatic (the increase of Ala, Gly, Val, Leu > the decrease of Ile, Ser, Asp, Glu) and imino amino acids (the increase of Pro > the decrease of Hyp) also increased from a combinative view. As for the decrease in hydrophilic and the increase in hydrophobic amino acids, it was speculated that the latter were surrounded by the former when the fresh proteins were in liquid states. With long-term dryness and aging, the hydrophilic amino acids outside were degraded so that the inner hydrophobic amino acids were exposed gradually [26]. Meanwhile, the decreasing amounts of the negative-charged Asp and Glu were higher than those of the positive-charged Lys. Hence, the original charge distribution was disturbed, which harmed the stability of the proteins.

Of particular interest was that the contents of those unstable amino acids Phe, Met, and Tyr in collagen were rather low compared with egg and casein. Hence, collagen's ability to resist the natural aging process was best due to it having the highest amount of stable amino acids [27]. Other research has also pointed out that the appearance of collagen did not change obviously after light aging and it behaved the best towards natural aging [2,4,28].

The Aging-Accelerating Role of Pigment

A previous study has shown that metal ions could induce ligand–receptor reaction and change the structure and activity of proteins by acting as ligands with atoms such as S, O, and N. They then sped up the aging process and resulted in the fracture of silk fiber and a reduction in strength [29]. Karpowicz also indicated that pigment could promote aging since it was an important sensitizer [26]. As a result, it was reasonable that the contents of amino acids in the mixed binders changed more obviously than those of the corresponding pure binders.

3.2. FTIR Results

3.2.1. Spectral Changes

Figure 3 shows the FTIR comparative results of pig glue, whole egg, egg white, egg yolk, and casein before and after natural aging. The absorbance characteristics are specified in Table 3 [30–32]. It could be speculated that: (1) The consistent stair-step pattern formed by Amide I (1633~1663 cm^{-1}), Amide II (1533~1562 cm^{-1}), and Amide III (1232~1242 cm^{-1}) of the five pure binders demonstrates that they still possessed the typical protein FTIR absorbance features [33]. However, the peaks in Amide II declined remarkably and those of Amide III disappeared in the mixed binders, suggesting that the mixed binders had lost the protein's FTIR characteristics and ultramarine could speed up aging. (2) The Amide A regions (3281~3451 cm^{-1}) all broadened and shifted. Doyle indicated that the broadening of Amide A illustrated that the helix structure had become dissociated [34]. Moreover, hydrogen bond, as one substantial factor for maintaining helix structure, was destroyed with aging, which would result in the shift of this absorbance [35,36]. (3) The lipid typical peaks (CH_2 asymmetric stretch, CH_2 symmetric stretch, and C=O stretch) in whole egg (2855~2925 cm^{-1}, 1745~1712 cm^{-1}) and egg yolk (2854~2925 cm^{-1}, 1746~1712 cm^{-1}) shifted and become weaker. Binders with ultramarine changed even more intensely, for instance, C=O in the whole egg mixed binder had completely gone, stressing the accelerated aging function of ultramarine. (4) The glucosidic bonds in whole egg and egg white showed evident changes; 1167, 1095 cm^{-1} became merged in 1198 cm^{-1} in whole egg, while 1159 and 1074 cm^{-1} showed blue shifts to 1184 cm^{-1} and 1076 cm^{-1}, which was consistent with an early study indicating that the glucosidic bond was closely related to the stability of proteins [37]. (5) Except for in the pure pig glue, there were marked enhancements in CH_3 symmetric bending vibration peaks (1377~1403 cm^{-1}), which were important signs of aging [38].

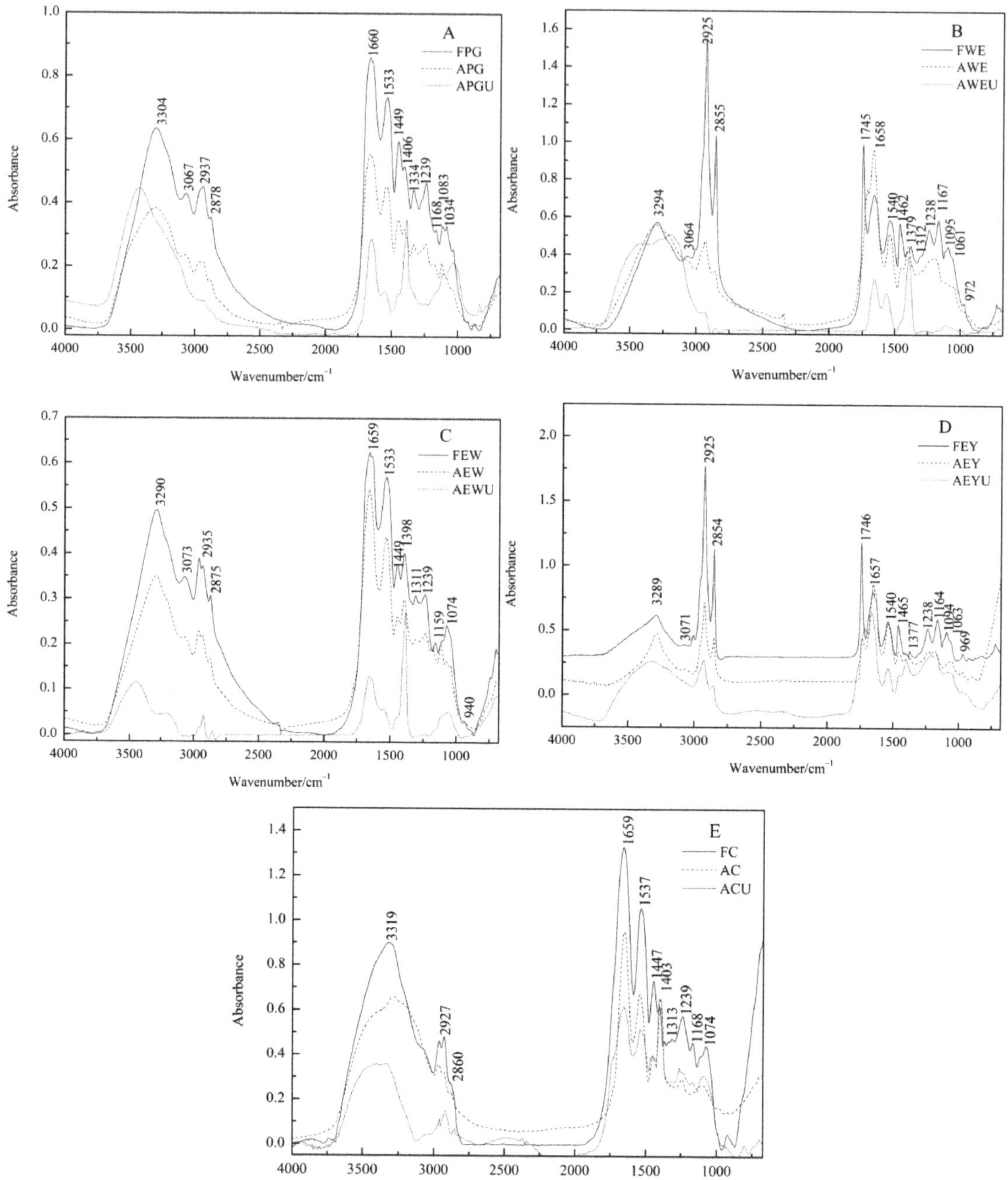

Figure 3. The FTIR spectra of the five fresh and aged binders ((**A**) pig glue; (**B**) whole egg; (**C**) egg white; (**D**) egg yolk; (**E**) casein).

Table 3. The absorbance characteristics of the five fresh and aged binders.

Region	Wavenumber/cm^{-1}															Characteristics
	FPG	APG	APGU	FWE	AWE	AWEU	FEW	AEW	AEWU	FEY	AEY	AEYU	FC	AC	ACU	
Amide A	3304	3306	3436	3294	3301	3281	3290	3293	3451	3289	3284	3314	3319	3308	3286	N–H stretch, O–H stretch
Amide B	3067	3074	/	3064	3071	/	3073	3078	3211	3071	/	/	/	/	/	C–N stretch
	2937	2939	2931	2925	2930	2940	2935	2935	2926	2925	2925	2929	2927	2921	2932	CH_2 asymmetric stretch
	2878	2879	2850	2855	2858	2855	2875	2876	2854	2854	2855	2857	2860	2852	2878	CH_2 symmetric stretch
	/	/	/	1745	1712	/	/	/	/	1746	1737	1712	/	/	/	C=O stretch
Amide I	1660	1660	1651	1658	1633	1656	1659	1660	1657	1657	1654	1663	1659	1655	1653	C=O stretch
Amide II	1533	1536	1550	1540	1537	1562	1533	1533	1548	1540	1542	1539	1537	1538	1543	N–H bend, C–N stretch
	1449	1450	1450	1462	1449	/	1449	1449	1457	1465	1459	1454	1447	1458	1447	CH_3 asymmetric bend
	1406	1406	1386	1379	1400	1396	1398	1400	1387	1377	1379	1400	1403	1400	1400	CH_3 symmetric bend
Amide III	1334	1334	1274	1312	/	/	1311	1308	/	/	/	/	1313	1312	1316	CH_2 in-plane bend
	1239	1243	1241	1238	/	/	1239	1237	1238	1238	1232	/	1239	1232	1242	N–H bend
	1165	1182	/	1167	1198	/	1159	1184		1164	1171	1185	1168	1166	1163	Ala, Glucoside
	1083	1084	1109	1095	/	1109	1074	1076	1074	1094	1083	/	1074	1091	1097	C–O stretch/C–N–C stretch, Glucoside
	1034	1037	1033	1061	1047	/	/	/	/	1063	1059	1059	/	/	/	Gly–Gly peptide chain
	/	/	/	972	968	/	940	945	/	969	969	972	/	/	/	Gly–Ala peptide chain
	874	875	/	/	/	/	/	/	/	/	/	/	/	/	/	C–C stretch of Hyp

3.2.2. Contents of the Secondary Structures

The attributions and contents of the secondary structures [39–41] are illustrated in Table 4. By calculating the variations of the six secondary structures (β-sheet = parallel β-sheet+ anti parallel β-sheet), it could be concluded that: (1) the contents of α-helix dramatically decreased, with the minimum reduction exceeding 20% and a maximum of 40.67% (AEWU); (2) the contents of β-sheet showed increasing trends with an average increment of 16.63%; (3) the contents of random coils climbed by an average increasing amount of 14.06% with a maximum of 23.39% (AEWU); (4) β-turns decreased at a range of −9.63%~−2.20%; (5) the contents of the secondary structures in the mixed binders changed more than those in the pure binders.

Table 4. The attributions and contents of the five secondary structures of the five fresh and aged binders.

Binders	Secondary Structures	Fresh		Aged Pure		Aged Mixture	
		Wavenumber /cm^{-1}	Contents /%	Wavenumber /cm^{-1}	Contents /%	Wavenumber /cm^{-1}	Contents /%
Pig glue	parallel β-sheet	1631.08	19.63	1629.18	18.23	1623.21	15.18
	random coils	1643.13	2.77	1643.12	12.53	1643.15	15.66
	α-helix	1658.79	41.82	1657.5	19.58	1662.97	17.23
	anti parallel β-sheet	1683.32	31.35	1678.93	47.42	1674.71	50.12
	β-turns	1695.76	4.43	1694.95	2.23	1691.18	1.81
Whole egg	parallel β-sheet	1629.58	8.32	1620.51	9.53	1626.86	10.33
	random coils	1643.59	12.72	1634.7	26.65	1642.66	28.64
	α-helix	1660.35	50.05	1654.23	25.96	1657.2	22.98
	anti parallel β-sheet	1675.24	18.19	1671.91	32.48	1675.15	36.96
	β-turns	1693.79	10.71	1690.76	5.39	1692.85	1.08
Egg white	parallel β-sheet	1629.53	7.81	1623.85	6.07	1630.13	6.95
	random coils	1647.29	8.34	1639.88	29.98	1645.43	31.74
	α-helix	1669.07	59.2	1663.52	24.14	1666.51	18.53
	anti parallel β-sheet	1686.44	17.21	1679.37	37.04	1679.83	41.05
	β-turns	1699.67	7.44	1689.96	2.76	1692.75	1.73
Egg yolk	parallel β-sheet	1624.15	9.26	1622.18	13.43	1628.49	12.1
	random coils	1635.07	14.88	1637.29	24.98	1644.04	26.45
	α-helix	1651.5	40.27	1658.87	19.58	1661.85	15.86
	anti parallel β-sheet	1671.88	20.21	1675.64	29.34	1677.49	32.75
	β-turns	1693.18	15.38	1690.36	12.67	1692.8	12.83
Casein	parallel β-sheet	1621.75	16.74	1626.9	13.84	1625.78	11.8
	random coils	1639.73	15.54	1645.83	26.93	1641.69	25.51
	α-helix	1659.57	27.93	1662.12	5.31	1662.04	5.23
	anti parallel β-sheet	1674.68	29.48	1676.13	46.82	1676.04	51.21
	β-turns	1693.32	10.31	1688.4	7.1	1686.74	6.25

Large amounts of research have proven that light and thermal aging would result in the decrease of α-helix, the increase of β-sheet, and random coils [42–45]. This signifies that the ordered structure of proteins transforms into disordered states, partially because the bonds' vibrations are intensive, the stretching and bending energy are increasing, and the spatial structure becomes loose with the increase in temperature. The reasons why the secondary structures showed the above changes were various. (1) α-helix would transform into β-sheet and random coils since the entropy was climbing [46,47]. (2) α-helix was more likely to be degraded because it located in the vulnerable amorphous region [48], while β-sheet was non vulnerable since it was in the stable crystalline region [49]. (3) It was confirmed that Pro impeded α-helix [50,51], while Pro, Gly, Asp, and Ser were applied more in β-turns [52,53]. Hence, α-helix decreased due to the climb of Pro while β-turns were lowered, probably attributed to the reduction of Asp. Ser exerted more influence than the increase of Pro and Gly. (4) The interactive function between protein and metal ions could change conformations. For instance, cinnabar (HgS) could form stable complexes with proteins and act as a sensitizer in cross-linking, hydrolysis, and oxidation [54]. Several pigments were known to catalyze binder deterioration [55] while finer grained pigments could provoke organic–inorganic interaction [56]. Hence, crosslinking would be induced, which would destroy the original spatial structure. It was thus speculated that the contents of the secondary structures of the mixed binders varied more.

4. Conclusions

After 10 years' natural aging, the energy and free particles produced in reactions resulted in the breaking down of the unstable Met, Lys, Ile, Ser, Asp, Glu, and Hyp (especially aromatic Phe and Tyr), and thus a slight rise in the stable Ala, Gly, Val, Leu, and Pro in the five proteinaceous binders. The unbalanced distribution of the positive and negative charged amino acids, the broadening of Amide A, the decrease of α-helix, and the increase of β-sheet and random coils all illustrated that the ordered structures of the proteins had experienced deteriorations and transformed into disordered states.

Due to the small amount of the unstable Phe, Tyr, Met, and Lys, pig glue had better resistance towards natural aging with less content fluctuation of amino acids and secondary structures, compared with egg and casein. Ultramarine could accelerate the process of natural aging, revealing, by evident variations in the contents of amino acids and secondary structures, FTIR spectra changes and the appearance of CH_3 symmetric bending vibrations. It was also found that lipids in egg were more likely to be degraded during the aging process.

Author Contributions: Conceptualization, Z.M.; methodology, L.W.; software, L.Y.; validation, L.W.; formal analysis, Z.M.; investigation, T.B.; resources, V.P.; data curation, L.Y.; writing—original draft preparation, Z.M.; writing—review and editing, L.W.; visualization, Z.M.; supervision, X.Z.; project administration, G.K.; funding acquisition, X.Z. All authors have read and agreed to the published version of the manuscript.

Funding: This research was funded by [National Natural Science Foundation of China] grant number [No. 52072228], [National Key R&D Program of China] grant number [No. 2019YFC1520100], [China Scholarship Council] grant number [No. 201506970013] and [Shaanxi Key Industry Innovation Chain Project] grant number [No. 2021ZDLSF06-05].

Institutional Review Board Statement: Not applicable.

Informed Consent Statement: Not applicable.

Data Availability Statement: Not applicable.

Conflicts of Interest: The authors declare no conflict of interest.

References

1. Ma, Z.Z.; Wang, L.Q.; Yang, L.; Zhao, X.C. The influence of the soil aging exerting on the stability of proteinaceous binders in Chinese polychromy artworks. *Microchem. J.* **2020**, *157*, 104955. [CrossRef]
2. Colombini, M.P.; Modugno, F.; Menicagli, E.; Fuoco, R.; Giacomelli, A. GC-MS characterization of proteinaceous and lipid binders in UV aged polychrome artifacts. *Microchem. J.* **2000**, *67*, 291–300. [CrossRef]
3. Colombini, M.P.; Modugno, F.; Fuoco, R.; Tognazzi, A. A GC-MS study on the deterioration of lipidic paint binders. *Microchem. J.* **2002**, *73*, 175–185. [CrossRef]
4. Wei, S.Y. *A Study of Ancient Natural Organic Binding Media*; University of Science and Technology: Vienna, Austria, 2007.
5. Wang, Z.M.; Wang, L.Q.; Ma, Z.Z.; Yang, L. Methodological study of proteinaceous binders in artworks by GC-MS. *J. Chin. Mass. Spectrom. Soc.* **2019**, *40*, 335–341.
6. Ma, Z.Z.; Yan, J.; Zhao, X.C.; Wang, L.Q.; Yang, L. Multi-analytical study of the suspected binding medium residues of wall paintings excavated in Tang tomb, China. *J. Cult. Herit.* **2017**, *24*, 171–174. [CrossRef]
7. Shen, R.Q.; Gu, Q.M. *Basic Biochemistry*; Shanghai Scientific Technology Press: Shanghai, China, 1980.
8. Yang, H.L.; Jiang, X.N. *Basic Biochemistry*; China Forestry Press: Beijing, China, 2015.
9. Zhang, X.M.; Yuan, S.X. Research on the evaluation of the deterioration degree of silk fabrics. *Sci. Conserv. Archaeol.* **2003**, *15*, 31–37.
10. Song, Q.S.; Sun, S.X. *Inorganic Chemistry Course*; Shandong University Press: Jinan, China, 2001.
11. Zhang, H.M.; Wu, X.Y.; Fen, Y.F. Primarily probing into photodegradation mechanism of mulberry. *J. Wuhan. Text. I.* **1997**, *10*, 1–5.
12. Davies, M.J. The oxidative environment and protein damage. *BBA* **2005**, *1703*, 93–109. [CrossRef]
13. Shao, H.; Wang, S.L.; You, Z.Y.; Li, A. Oxygen free radicals and protein metabolism. *For. Med. Mol. Bio.* **1990**, *12*, 42–44.
14. Wang, Z.Q. *Chemical Modification of Tyrosine residues of Silk Protein and the Effect on Its Light Stabilization*; Zhejiang Sci. Tech University: Hangzhou, China, 2016.
15. Baltova, S.; Vassileva, V. Photochemical behaviour of natural silk-II.Mechanism of fibroin photodestruction. *Polym. Degrad. Stab.* **1998**, *60*, 61–65. [CrossRef]
16. Millington, K.R. Photoyellowing of wool. Part 1 Factors affecting photoyellowing and experimental techniques. *Color. Technol.* **2006**, *122*, 169–186. [CrossRef]
17. Wang, Z.Q.; Li, J.J.; Zhang, H.L. Influence of benzotriazole UV absorber on photostability of silk. *J. Text. Res.* **2016**, *37*, 109–114.
18. Xu, C.G.; Wang, Z.Q.; Cui, Z.H.; He, K.J.; Chen, W.G. Research of effects of UV absorber on improving photostablity of amino acids. *J. Zhejiang. Sci. Tech. Uni.* **2014**, *31*, 112–132.
19. Bringans, S.D.; Dyer, J.M.; Plowman, J.E. Kynurenine located within keratin proteins isolated from photoyellowed wool fabric. *Text. Res. J.* **2006**, *76*, 288–294. [CrossRef]
20. Barbabietola, N.; Tasso, F.; Alisi, C.; Marconi, P.; Perito, B.; Pasquariello, G.; Sprocati, A.R. A safe microbe-based procedure for a gentle removal of aged animal glues from ancient paper. *Int. Biodeter. Biodeg.* **2016**, *109*, 53–60. [CrossRef]
21. Jaenicke, R. Stability and stabilization of globular proteins in solution. *J. Biotechnol.* **2000**, *79*, 193–203. [CrossRef]
22. Zhang, X.M.; Yuan, S.X. Analytical research of the amino acids of aging silk. *Sci. Conserv. Archaeol.* **2003**, *15*, 18–203.
23. Song, R.R.; Bao, B.; Bu, Y.S.; Wang, Y.X.; Chen, L.J.; Wu, W.H. Studies on the thermostability, circular dichroism and Infrared spectral characteristics of type II collagen. *Chin. J. Mar. Drugs.* **2013**, *32*, 55–62.
24. Wang, F.F.; Fu, J.Q. Study on damage of artificially aged tussah silk fiber. *J. Beijing. I. Cloth. Technol.* **2008**, *28*, 32–36.
25. Lu, C.B.; Huang, S.Z.; Fu, J.R. Relation of heat stability of storage proteins and amino acid composition in peanut seeds. *Seed* **2001**, *3*, 3–5.
26. Karpowicz, A. Aging and deterioration of proteinaceous media. *Stud. Conserv.* **1981**, *26*, 153–160.
27. Spikes, J.D.; Livingston, R. The molecular biology of photodynamic action: Sensitized photooxidation in biological systems. *Adv. Radiat. Biol.* **1969**, *3*, 29–121.
28. Sionkowska, A. The inflfluence of UV light on collagen/poly(ethylene glycol) blends. *Poly. Degrad. Stab.* **2006**, *91*, 305–312. [CrossRef]
29. Zhang, X.M.; Zong, X.L.; Yang, Y.L. The effect of Fe and Cu on the conservation state of silk textiles. *Sci. Conserv. Archaeol.* **2010**, *22*, 1–8.
30. Pellegrini, D.; Duce, C.; Bonaduce, I.; Biagi, S.; Ghezzi, L.; Colombini, M.P.; Tine, M.R.; Bramanti, E. Fourier transform infrared spectroscopic study of rabbit glue inorganic pigments mixtures in fresh and aged reference paint reconstructions. *Microchem. J.* **2015**, *124*, 31–35. [CrossRef]
31. Sotiropoulou, S.; Papliaka, Z.E.; Vaccari, L. Micro FTIR imaging for the investigation of deteriorated organic binders in wall painting stratigraphies of different techniques and period. *Microchem. J.* **2016**, *124*, 559–567. [CrossRef]
32. Orsini, S.; Bramanti, E.; Bonaduce, I. Analytical pyrolysis to gain insights into the protein structure, the case of ovalbumin. *J. Anal. Appl. Pyrolysis.* **2018**, *133*, 59–67. [CrossRef]
33. Derrick, M.R.; Stulik, D.; Landry, J.M. *Infrared Spectroscopy in Conservation Science*; J-Paul Getty Trust: Los Angeles, CA, USA, 1999.
34. Ma, Z.Z.; Wang, L.Q.; Krist, G.; Bayerova, T.; Yang, L. Study on secondary structural changes of proteinaceous binding media in ancient polychromy artwork after Light aging by FTIR spectroscopy. *Spectrosc. Spectra. Anal.* **2017**, *37*, 2712–2716.
35. Sionkowska, A. Effects of solar radiation on collagen and chitosan films. *J. Photochem. Photobiol. B* **2006**, *82*, 9–15. [CrossRef]

36. Ren, G.D.; Guo, A.L.; Geng, F.; Ma, M.H.; Huang, Q.; Wu, X.F. Study of the effect of temperature on the conformation of ovotransferrin by two-dimensional infrared correlation spectroscopy. *Spectrosc. Spectra. Anal.* **2012**, *32*, 1780–1784.
37. Shan, Y.Y.; Ma, M.H.; Huang, Q.; Gao, F. Infrared spectroscopy analysis of structural changes of ovornucin as induced by temperature. *Chem. J. Chin. Uni.* **2012**, *33*, 1950–1956.
38. Yuan, M.M.; Chen, C.J.; Yue, J.; Wang, G.H. Analysis of changes in protein molecular structure of ancient silk based on infrared spectrum. *J. Silk.* **2013**, *50*, 7–10.
39. Sazonova, S.; Grube, M.; Shvirksts, K.; Galoburda, R.; Gramatina, I. FTIR spectroscopy studies of high pressure-induced changes in pork macromolecular structure. *J. Mol. Struct.* **2019**, *1186*, 377–383. [CrossRef]
40. Baltacıoğlu, H.; Bayındırlı, A.; Severcan, F. Secondary structure and conformational change of mushroom polyphenol oxidase during thermosonication treatment by using FTIR spectroscopy. *Food. Chem.* **2017**, *214*, 507–514. [CrossRef]
41. Ulrichs, T.; Droteff, A.M.; Ternes, W. Determination of heat-induced changes in the protein secondary structure of reconstituted livetins (water-soluble proteins from hen's egg yolk) by FTIR. *Food. Chem.* **2015**, *172*, 909–920. [CrossRef]
42. Badillo-Sanchez, D.; Chelazzi, D.; Giorgi, R.; Cincinelli, A.; Baglioni, P. Characterization of the secondary structure of degummed Bombyx mori silk in modern and historical samples. *Poly. Degrad. Stab.* **2018**, *157*, 53–62. [CrossRef]
43. Fu, J.; Zhao, L.L.; Wang, J.H. Conformation transformation of Aβ42 protein under different temperature by molecular dynamics simulations. *Hans. J. Biomed.* **2013**, *3*, 1–6. [CrossRef]
44. Tankovskaia, S.A.; Abrosimova, K.V.; Paston, S.V. Spectral demonstration of structural transitions in albumins. *J. Mol. Struct.* **2018**, *1171*, 243–252. [CrossRef]
45. Duconseille, A.; Wien, F.; Audonnet, F.; Traore, A.; Refregiers, M.; Astruc, T.; Sante-Lhoutellier, V. The effect of origin of the gelatine and ageing on the secondary structure and water dissolution. *Food Hydrocolloid.* **2017**, *66*, 378–388. [CrossRef]
46. Zhao, Q.; Buehler, M.J. Molecular dynamics simulation of the -helix to -sheet transition in coiled protein filaments evidence for a critical filament length scale. *Physic. Rev. Lett.* **2010**, *104*, 198304.
47. Ding, F.; Borreguero, J.M.; Buldyrey, S.V.; Stanley, H.E.; Dokholyan, N.V. Mechanism for the -helix to -hairpin transition. *Proteins* **2003**, *53*, 220–228. [CrossRef]
48. Hu, X.; Kaplan, D.; Cebe, P. Dynamic protein-water relationships during β-Sheet formation. *Macromolecules* **2008**, *41*, 3939–3948. [CrossRef]
49. Wang, J.N.; Lu, C.D. Study on β-Sheet structure formed by self-assembly of fibroin crystalline typical peptides using thioflavine T fluorescence probe. *Acta. Chim. Sinica.* **2007**, *65*, 111–115.
50. Wen, J.K. *Biochemistry*; Traditional Chinese Medicine Press: Beijing, China, 2008.
51. Lin, H.; Guo, F.B.; Wang, D. *Concise Bioinformatics*; Press of Electronic Science and Technology University: Chengdu, China, 2014.
52. Guan, B.; Lin, H.; Wang, G.C. *Food Protein Chemistry*; Chemical Industry Press: Beijing, China, 2005.
53. Wang, L.J. *Principles of Biochemistry and Molecular Biology*; Science Press: Beijing, China, 1999.
54. Duce, C.; Ghezzi, L.; Onor, M.; Bonaduce, I.; Colombini, M.P.; Tine, M.R.; Bramanti, E. Physico-chemical characterization of protein-pigment interactions in tempera paint reconstructions casein cinnabar and albumin cinnabar. *Anal. Bioanal. Chem.* **2012**, *402*, 2183–2193. [CrossRef]
55. Anaf, W.; Schalm, O.; Janssens, K.; Wael, K.D. Understanding the (in)stability of semiconductor pigments by a thermodynamic approach. *Dyes Pigments* **2015**, *113*, 409–415. [CrossRef]
56. Elert, K.; Cardell, C. Weathering behavior of cinnabar-based tempera paints upon natural and accelerated aging. *Spectrochim. Acta A* **2019**, *216*, 236–248. [CrossRef]

coatings

MDPI

Article

The Process and Mechanisms of Freezing and Thawing Deterioration of Petroglyphs under Snowmelt and Rainfall Infiltration

Guopeng Wu, Kai Cui *, Chengrui Ge and Xiangpeng Yu

Key Laboratory of Disaster Prevention and Mitigation in Civil Engineering of Gansu Province, Department of Geotechnical Engineering, Lanzhou University of Technology, 287 Langongping Road, Lanzhou 730050, China
* Correspondence: cuik09@lut.edu.cn

Abstract: Freeze–thaw cycles under the conditions of snowmelt and rainfall infiltration in winter are triggers for petroglyph weathering. Rock samples from the Helan Mountains were subjected freeze–thaw cycles under various conditions in laboratory testing, and the mass, elastic wave velocity and unconfined compressive strength were tested. The results show that the mass loss rate, wave velocity and compressive strength decreased. In addition, according to XRD analyses, the content of calcite and feldspar in the samples decreased, and according to SEM analyses, the microscopic pores grew. Considering the mass, mineral content and micro-voids, the variation decreases in the order: snowmelt group, rainfall infiltration group, normal freeze–thaw group. However, for wave velocity and compressive strength, the opposite results were observed. This research contributes to a better understanding of the weathering processes and mechanisms of petroglyphs in the Helan Mountains in winter.

Keywords: petroglyph; weathering; snow melting; macro-property; micro-structure

1. Introduction

Over 5500 single petroglyphs have been found on the slopes and floodplain fans outside Helan Pass in Ningxia's Helan Mountains (Figure 1a), with the themes of animals, plants, human bodies, human faces and symbols (Figure 1b). Many of the petroglyphs incised in the rock of the Helan Mountains are famous for their rich expression and profound cultural connotations [1]. However, some deterioration such as erosion, flake buckling, peeling etc. occurs and the rock is seriously weathered when exposed to the open air for a long time [2]. Especially in recent years, due to drastic environmental changes, the weathering process of the rocks has accelerated, leading to the danger that the petroglyphs will disappear. Therefore, it has become an urgent issue to study the weathering mechanisms of petroglyph carriers caused by environmental factors and formulate scientific conservation measures. In addition, the unique climatic characteristics of Helankou mean that snowmelt, rainfall infiltration and common freeze–thaw weathering are important factors that cannot be ignored in the process of rock weathering of petroglyphs [3].

A lot of research has been carried out on the effects on rocks under the action of freeze–thaw cycles, and informative research results have been obtained. It is universally accepted that temperature, moisture content and rock type are the main factors causing rock damage [4–6]. The importance of temperature conditions, such as minimum temperature and cooling rate, have been investigated in a number of studies [7], and moisture content has also been reported as a significant factor leading to rock failure following F–T cycles [8]. Rock properties including initial porosity, pore size distribution and mineral content also affect the likelihood of rock failure [9–11]. Nicholson [12] conducted a comparative analysis of the mass changes in different sedimentary rocks under different freeze–thaw cycles, and the results showed that the mass loss was greater with increasing numbers of cycles and relatively smaller for sedimentary rocks with high strength. Tan [13] studied the freeze–thaw cycles of sandstone in the temperature range of −40 °C to 40 °C and concluded that the compressive strength of sandstone deteriorates with an increase in the number of cycles. Zhang Jizhou [14] conducted a systematic experimental study on the freeze–thaw damage mechanisms and mechanical characteristics of specimens after freeze–thaw cycling for three types of rocks (mudstone, gabbro and dolomite), and the results showed that rock freeze–thaw damage is influenced by multiple factors, among which environmental factors are significant. Keping Zhou [15] et al. conducted a comparative study on the macroscopic properties and microscopic characteristics of granite after freeze–thaw cycles. They concluded that the physical properties of granite gradually decreased with the increase in cycles. The granite changed from medium to strongly weathered with the increase in freeze–thaw cycles because the internal pores of the rock were continuously damaged and expanded and the porosity increased [16].

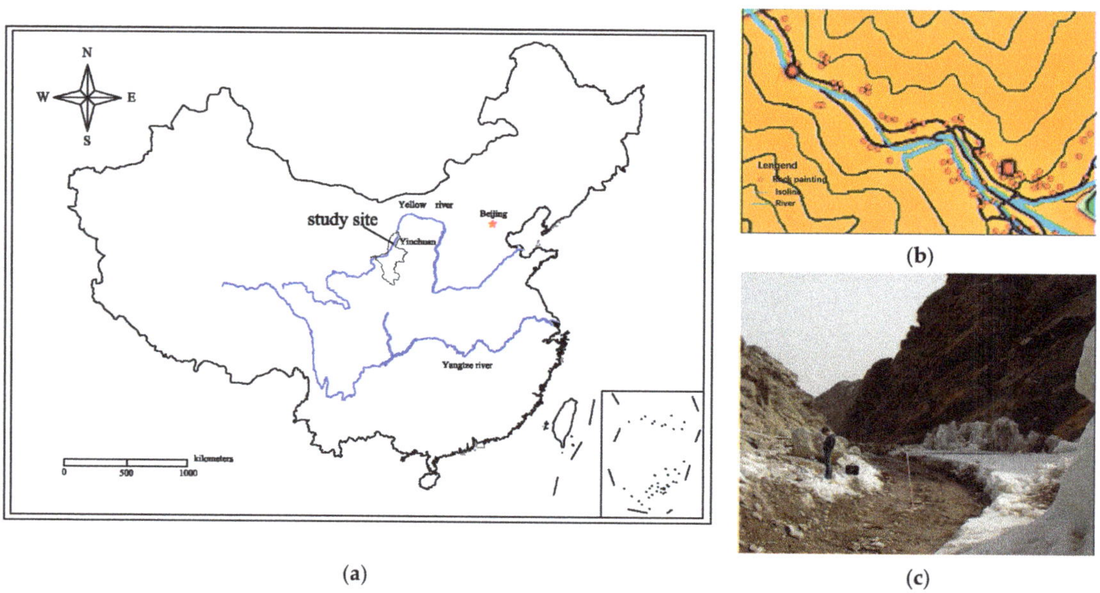

Figure 1. Location of the Helan Mountains petroglyphs: (**a**) the study site; (**b**) distribution of petroglyphs at study site; and (**c**) snow at study site.

In recent years, some scholars have considered the impact of snowfall on earth sites and rock sites. Pu [17] pointed out that the coupled effect of freeze–thaw cycles and salinity due to snowfall is a highly important driver for the development of soil sites in Qinghai-Tibet. Cui [18] simulated the freeze–thaw cycle process in soils in different temperature ranges in indoor tests. Cui [19] also study the deterioration mechanisms of rock art in the Helan Mountains under freeze–thaw cycles and different salinity conditions. Previous studies did not take into account the fact that the water recharge of snow into rock is slow as a result of the melting of snow, and does not occur uniformly throughout the snow layer [20]. The process of water infiltrating into the soil is slow with the snow melting, and the water added into the soil is not uniform or immediate [21]. The contact surface between the snow layer and the rock also produces a freezing-sticky effect, which affects the rock's integrity. This effect is a unique property of the overlying snow, which should also be a major focus for study [22].

In summary, scholars have studied the damage processes and mechanisms of sandstone, mudstone, tuff and other sedimentary rocks under the action of freezing and thawing. The research ideas and methods are becoming more mature. However, the carriers of petroglyphs in the Helan Mountains are meta-sandstones, and the rock carriers located on the cliff face and on the slope of the flood fan experience different types of freezing and thawing during the long winters in the area. However, little research has been reported on the weathering of these petroglyphs. Based on previous studies, in this paper, freeze–thaw tests are carried out on rock specimens under snowmelt and rainfall infiltration considering the actual environmental conditions of the Helankou petroglyph carrier in the Helan Mountains. A combined qualitative and quantitative analysis of the cumulative damage mechanisms of the specimens is conducted to provide some reference for the formulation of scientific conservation measures for the Helankou petroglyphs.

2. Study Background

The geographical coordinates of the Helan Mountains are 38°21′ N~39°22′ N and 105°49′ E~106°42′ E. The Yinchuan Plain and Alxa Plateau are on the east and west sides, respectively, with an altitude of about 2000~3000 m. The area belongs to the moderate-temperate arid, semi-arid plateau climate zone. According to survey data, the annual average temperature of the Helan Mountains is −0.9 °C, and the temperature range is large, with the annual extreme temperature reaching 25 °C and the lowest being −33 °C. Precipitation occurs during 62% of the year, and the annual average evapotranspiration is 1230 mm. The winter lasts for 5 months with a frost-free period of about 122.6 days, and the annual average wind speed is 7.7 m/s.

In recent years, human activities and environmental changes have intensified, and the destruction of Helankou petroglyphs has accelerated, with flaking, buckling and peeling, resulting in shallow and blurred carvings (Figure 2). According to survey statistics, about 660 petroglyphs were distributed within the 570-m-long petroglyph key reserve from the mouth of Helankou to the south bank of "Shuiguan". Among them, 173 petroglyphs showed flaking, buckling and peeling, accounting for 26.2% of the petroglyphs in this area, and 419 petroglyphs showed shallow and blurry carving marks, accounting for 63.4% of the petroglyphs in this section [23]. In addition, related studies have shown that flaking, buckling and spalling are mainly caused by physical weathering, while the shallow layer of the indentation is mainly due to chemical weathering of the erosion by aqueous chemical solutions [24]. Flaking, buckling and spalling refer to the deformation and uplift of the weathered crust on the surface of the rock mass and separation from the host rock, which results in a cavity area between the host rock the weathered crust that has not yet completely peeled off. Weathering refers to the phenomenon that the notches of the rock painting gradually become lighter until they disappear.

Figure 2. The main weathering types of the Helan Mountains petroglyphs [25]: (**a**) distribution of petroglyph weathering in Helankou; (**b**) peeling; (**c**) erosion; and (**d**) line drawing of petroglyph corresponding to (**c**).

3. Samples and Methods

Thin section identification tests were carried out to understand the mineral composition, grain size, and structure of the samples. The results showed that the rock consists of clastic and filler, and the rock is modified by metamorphic recrystallization. The debris content of the rock sample is 81%, which is composed of 57% quartz and 24% feldspar. The filler portion is 19%, which is composed of 12% mud matrix, 4% siliceous material, and 3% calcium.

Fresh and complete rock samples were taken from the Helankou petroglyph area. According to the "Engineering Rock Mass Test Method Standard" [26], the specimen was processed into a square with 70.7 mm sides and polished smooth. First, the rock specimens with obvious defects or appearance differences were rejected. Then the RSM-SY5 intelligent acoustic detector was used to measure the wave velocity of the specimen and the specimens with similar wave speeds were selected. The specimens were dried at 105 °C for 48 h until a constant weight was reached. After cooling, the samples were equilibrated at room temperature and the dried samples were saturated using suitable vacuum venting equipment.

Group X (Figure 3) (snow melt) were samples covered with a snow layer weighing 53.5 g [22]. To avoid over-melting of the snow, the rock samples were placed in a 5 °C environment for 12 h before the freeze–thaw cycle began. Group Y (rainfall infiltration) were samples sprayed from above with snow water of equal quality. The Group Z samples were not covered with snow or rain. The tests were carried out in accordance with the Code for Rock Test of Water Conservancy and Hydropower Engineering (DL/T 5368-2007, China). The rock samples were repeatedly saturated in distilled water at ambient temperature and then frozen in a refrigerator at about −30 °C. Referring to the average winter temperatures of Helankou since 2010, −30 °C was selected as the freezing temperature, 5 °C as the melting temperature, and 24 h as the freeze–thaw cycle. The freeze–thaw time was 12 h.

(a)

(b)

Figure 3. Samples. (**a**) Results of thin section identification test. (Pl: Plagioclase, Q: Quartz, Si: siliceous). (**b**) Samples for melting snow treatment.

Each set of test samples was divided into 5 groups of 5 samples each, which were subjected to 0, 15, 30, 45, and 60 F–T cycles, respectively. The mass loss rate and wave propagation velocity of the sample were recorded, and uniaxial compression experiments were carried out to study the mechanisms of mechanical change. In addition, X-ray diffraction (XRD) and scanning electron microscopy (SEM) analyses were performed to investigate the weathering mechanisms, mineral changes and micromorphology in the samples subjected to snowmelt and rainfall infiltration.

4. Results

4.1. Changes in Mass

The mass test of the samples was based on the test methods of rock for highway engineering (JTG E41-2005, China) specifications. The samples subjected to the snow melting and rainfall infiltration showed more significant changes in mass (Figure 4). The mass loss rate of samples was determined using the formula given in JTG E41-2005 Part 5 (Test methods of rock for highway engineering, 2005) as follows:

$$\Delta m = \frac{m_s - m_f}{m_s} \tag{1}$$

where Δm is the mass loss rate (%); m_s is the quality of saturated specimens before the test (g); and m_f is the quality of saturated specimens after the test.

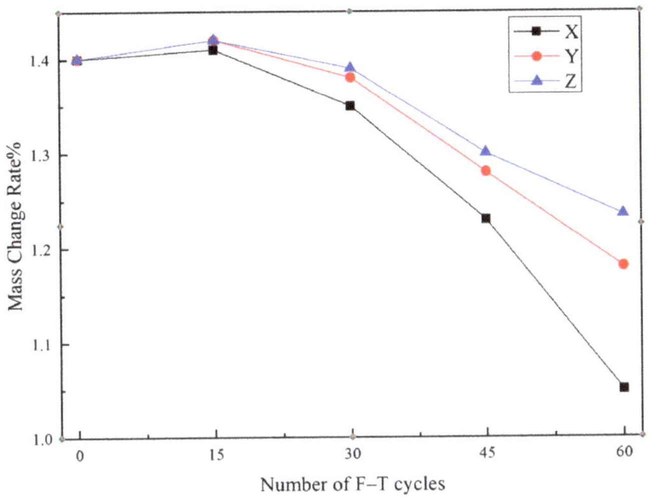

Figure 4. Mass change rate curve of three freeze–thaw cycle treatments.

As the weathering cycle progressed, the rate of rock mass loss increased slowly before the 15th cycle, with the treatment groups ordered as follows: Z, Y, X. Then, it decreased significantly after the 15th F–T cycle in the following order: X, Y, Z.

4.2. Changes in Wave Velocity

An RSM-SY5 intelligent acoustic detector was used to test the wave velocity of each group of samples. The results showed that wave velocity decreased in all experimental treatments (Figure 5). The wave velocity of the samples in Group X decreased by 21.44%, the samples in Group Y decreased by 19.89%, and the samples in Group Z decreased by 17.53%.

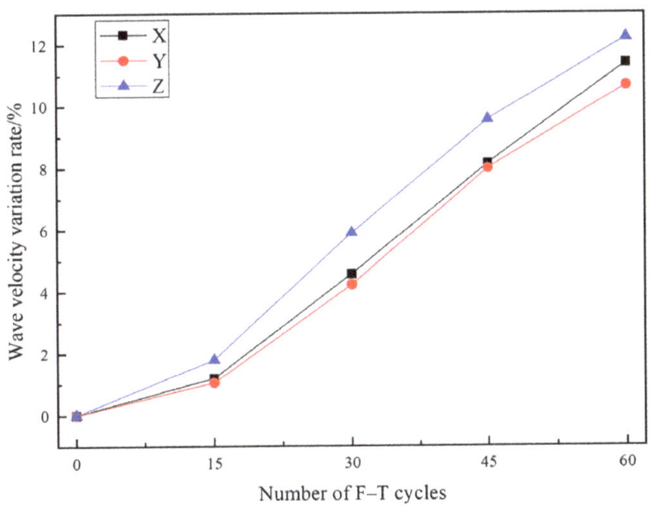

Figure 5. Wave velocity curve of three freeze–thaw cycle treatments.

4.3. Variation in Uniaxial Compression

In order to determine the strength of rock under snowmelt, rainfall infiltration and freeze–thaw treatments, uniaxial compression tests were performed and the coefficient of freezing resistance was calculated by Equation (2):

$$K_f = \frac{R_f}{R_s} \tag{2}$$

where K_f is the coefficient of freezing resistance; R_f is the compressive strength of the saturated specimens after F–T treatment (MPa); and R_s is the compressive strength of the saturated specimens before F–T treatment (MPa).

For Group X, the coefficient of freezing resistance decreased slowly in the freeze–thaw series, from 1.0 to 0.996 before the 15th cycle. After that, it changed faster, almost linearly, i.e., it decreased from 0.982 to 0.961 and, finally, to 0.931 by the 30th, 45th, and 60th cycles, respectively. Overall, the Group X sample strength decreased by 6.9%. For Group Y, the coefficient of freezing resistance decreased slowly before 30th cycle and quickly after that; overall, the Group Y sample strength decreased by 10.7%. However, unlike the X and Y groups, the change law of the frost resistance coefficient of Group Z almost conformed to the linear law, and decreased by 12.1% (Figure 6).

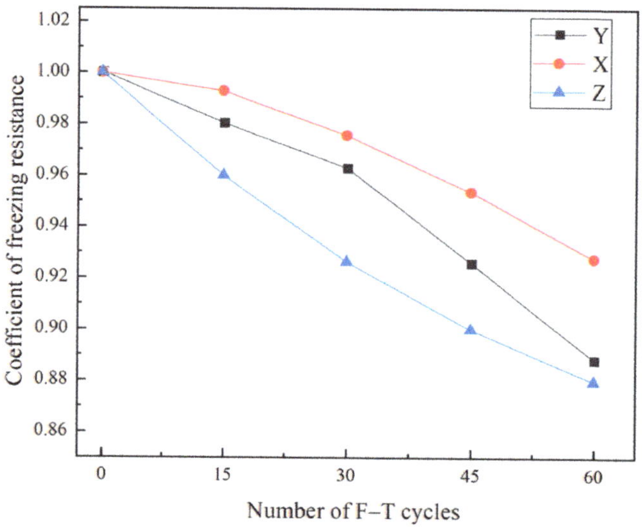

Figure 6. Compressive strength curve of three freeze–thaw cycle treatments.

4.4. Features of Stress–Strain Curves

After conducting uniaxial compression experiments on three different rock samples, it was revealed that the stress–strain curves of the specimens showed a common character. The four main stages of the process were defined as compaction, elastic deformation, yield and failure. The gradients of the stress–strain curves of all of the group samples declined, showing that their modulus of deformation decreased significantly in all three freeze–thaw cycle treatments. Moreover, the length of the compaction–deformation stage became longer with an increasing number of weathering cycles, whereas the elastic and yield stages were both shortened. Finally, the yield strength and ultimate strength of the samples decreased with an increasing number of weathering cycles (Figure 7).

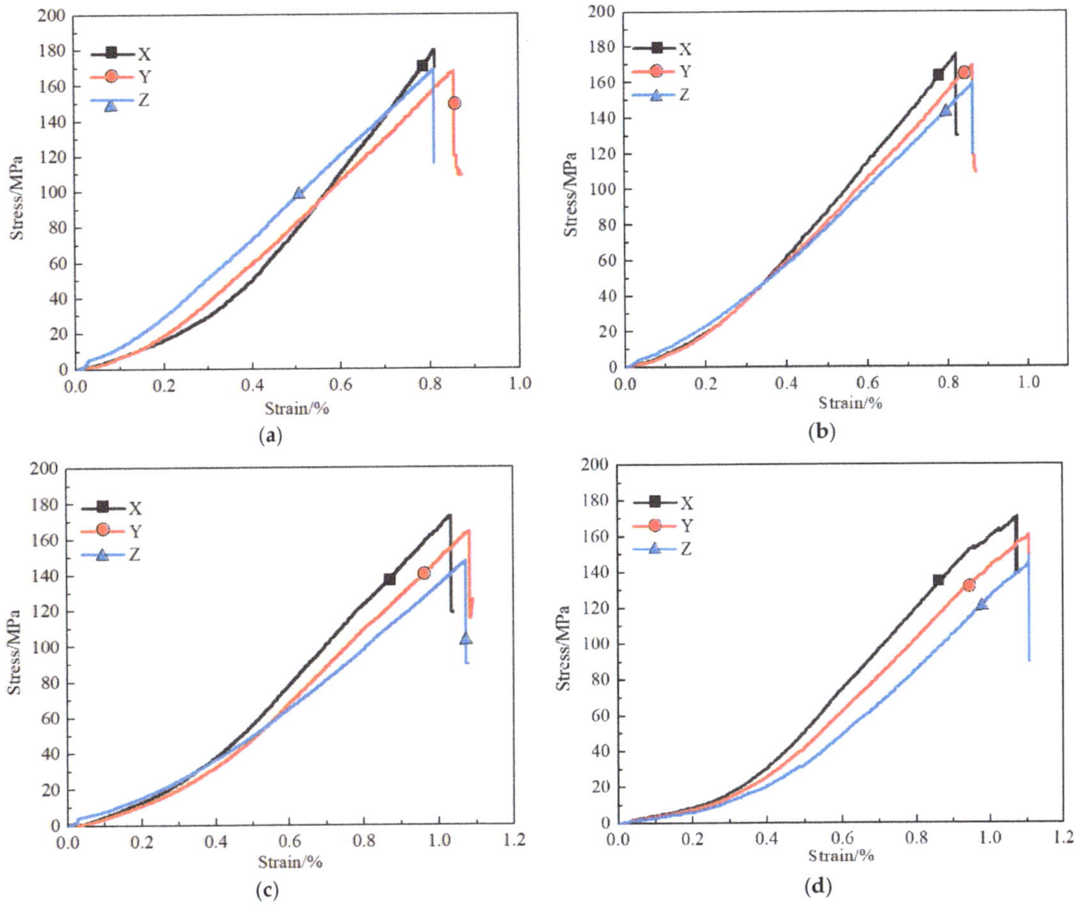

Figure 7. Stress–strain curve of three freeze–thaw cycle treatments at (**a**) 15 cycles; (**b**) 30 cycles; (**c**) 45 cycles; and (**d**) 60 cycles.

The stress–strain curves of the three treatment groups (snowmelt (X), rainfall infiltration (Y) and freeze–thaw cycles (Z)) showed obvious differences. In all experiments, the modulus of deformation, yield strength and ultimate strength decreased in the following order: X, Y, Z. The yield and failure stages of Group X exhibited distinct ductility features. Conversely, the Group Y samples showed no obviously distinct yield and failure stages, but did exhibit distinct rigidity features.

4.5. XRD Analysis

XRD analysis was conducted on rock samples of the petroglyphs to investigate changes in the mineralogical composition in the snow melting, rain infiltration and freeze–thaw treatments. A sample of 50 g was collected from the contact surface between the rock and snow, rock and rain and the arbitrary surface of rock in the freeze–thaw treatments, respectively. The samples were passed through a 0.075 mm sieve and different treatments (i.e., air-drying, heating and glycolation) were applied for the XRD tests. A Philips PW 3710 diffractometer was used for XRD analysis of the three slides. The diffraction patterns were determined using Cu–Kα radiation with a Bragg angle (2θ) range of 3–30°, running at a rate of 0.05°/s.

Figure 8 shows that quartz, clay, feldspars and calcite were the main minerals, with the clay containing kaolinite with a little illite. The mineral content of the unweathered sample is also shown in Figure 8b. Since quartz hardly reacts at low temperature, the results show that the content of clay increased while the content of feldspar, plagioclase, and calcite decreased during the snow melting, rain infiltration and freeze–thaw process. We also found the mineral content changed more dramatically in the snow melting group than in the others, e.g., the content of calcite decreased from 12% to 2% in the snow melting group versus 4% in the rain filtration group and up to 9% in the freeze–thaw group. In addition, the decrease in the calcite content was faster in the early stage of the snowmelt group and rain filtration group. It then became slower, and the changes in the feldspar and plagioclase content show the opposite trend.

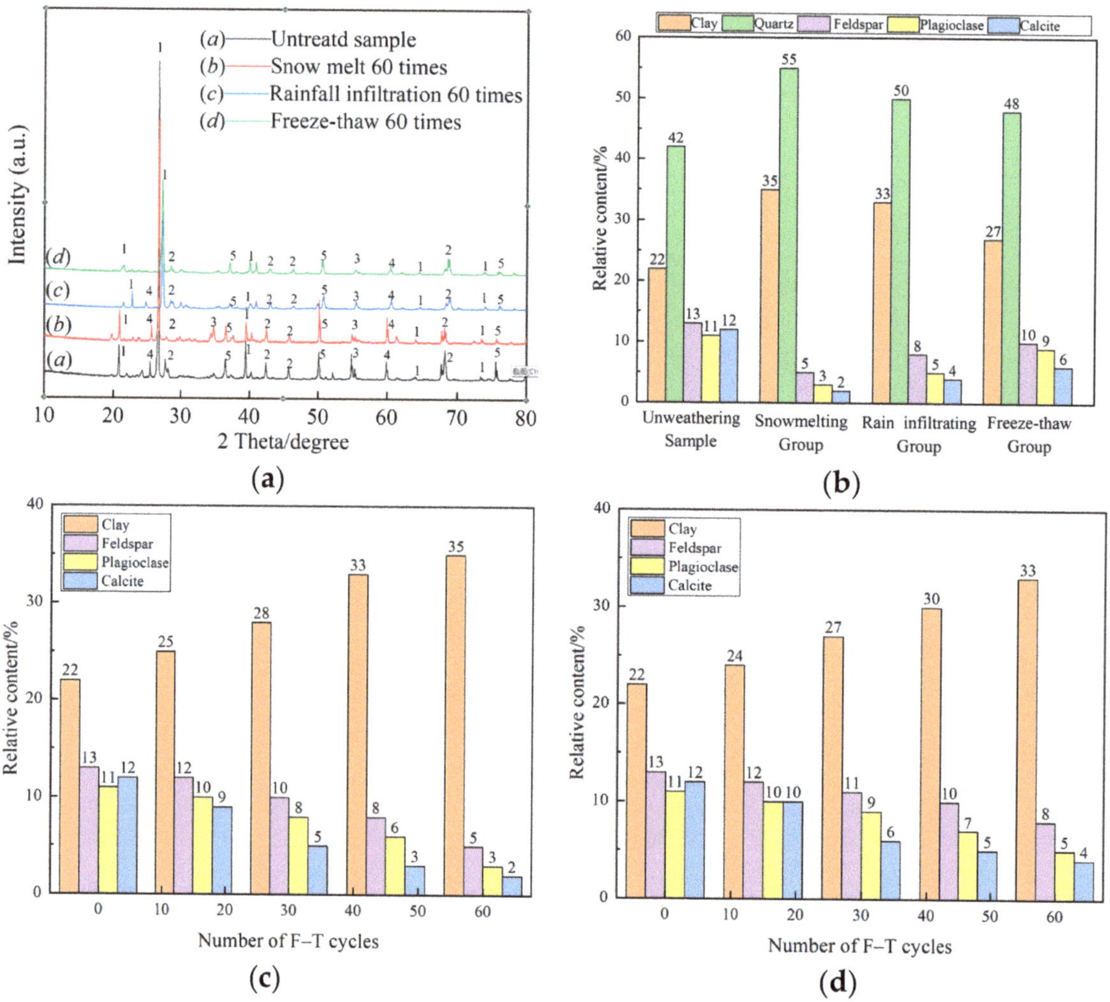

Figure 8. X-ray diffraction (XRD) scans of rock for three F–T cycle treatments. (**a**) XRD spectrogram (1—quartz, 2—clay, 3—feldspar, 4—plagioclase, 5—calcite). (**b**) Mineral content variation after three F–T cycle treatments. (**c**) Mineral content variation in snowmelt group. (**d**) Mineral content variation in rain infiltration group.

4.6. Microstructural Features

To investigate changes in the microstructure, SEM analysis was conducted on rock samples subjected to snow melting, rain infiltration and freeze–thaw treatments. The observation surfaces were selected on the contact surface between the rock and snow, and the rock and rain in the snow melting and rain infiltration treatments and on the arbitrary surface of the rock in the freeze–thaw treatment.

PCAS software of Nanjing University was applied in analysing the SEM images to calculate the particle size, porosity and pore width of samples.

Changes in the surface microstructure of the rock samples were manifested by the development, addition, expansion and interconnection of holes and linear gaps. The surface of the unweathered rock was smooth (Figure 9a). After snowmelting treatment, a large number of holes and increased linear porosity were found in the SEM images (Figure 9b). The rainwater filtration treatment led to increased porosity, but no linear porosity (Figure 9c). Compared with the test above, Figure 9d shows that the sample surface only became rough after the freeze–thaw treatment. In addition, evidence of calcite and feldspar corrosion and the increase and agglomeration of clay was found in the SEM images at 1000× magnification (Figure 9e–g). The parameters of porosity and pore width for the three freeze–thaw cycle treatments are detailed in Table 1.

Table 1. SEM parameters of the samples.

Groups	Number of Cycles	Particle Size (μm)	Porosity (%)	Pore Width (μm)
X	0	4–10	10–15	1–3
	15	6–10	12–20	2–6
	30	6–12	15–25	4–8
	45	10–18	15–25	8–12
	60	15–20	>25	>10
Y	0	4–10	10–15	1–3
	15	6–10	12–20	2–6
	30	6–10	12–20	2–6
	45	8–15	15–25	5–10
	60	10–18	15–25	8–12
Z	0	4–10	10–15	1–3
	15	6–10	10–15	1–3
	30	8–10	15–25	2–6
	45	10–15	15–25	2–6
	60	12–18	15–25	5–10

(a)

(b)

Figure 9. *Cont.*

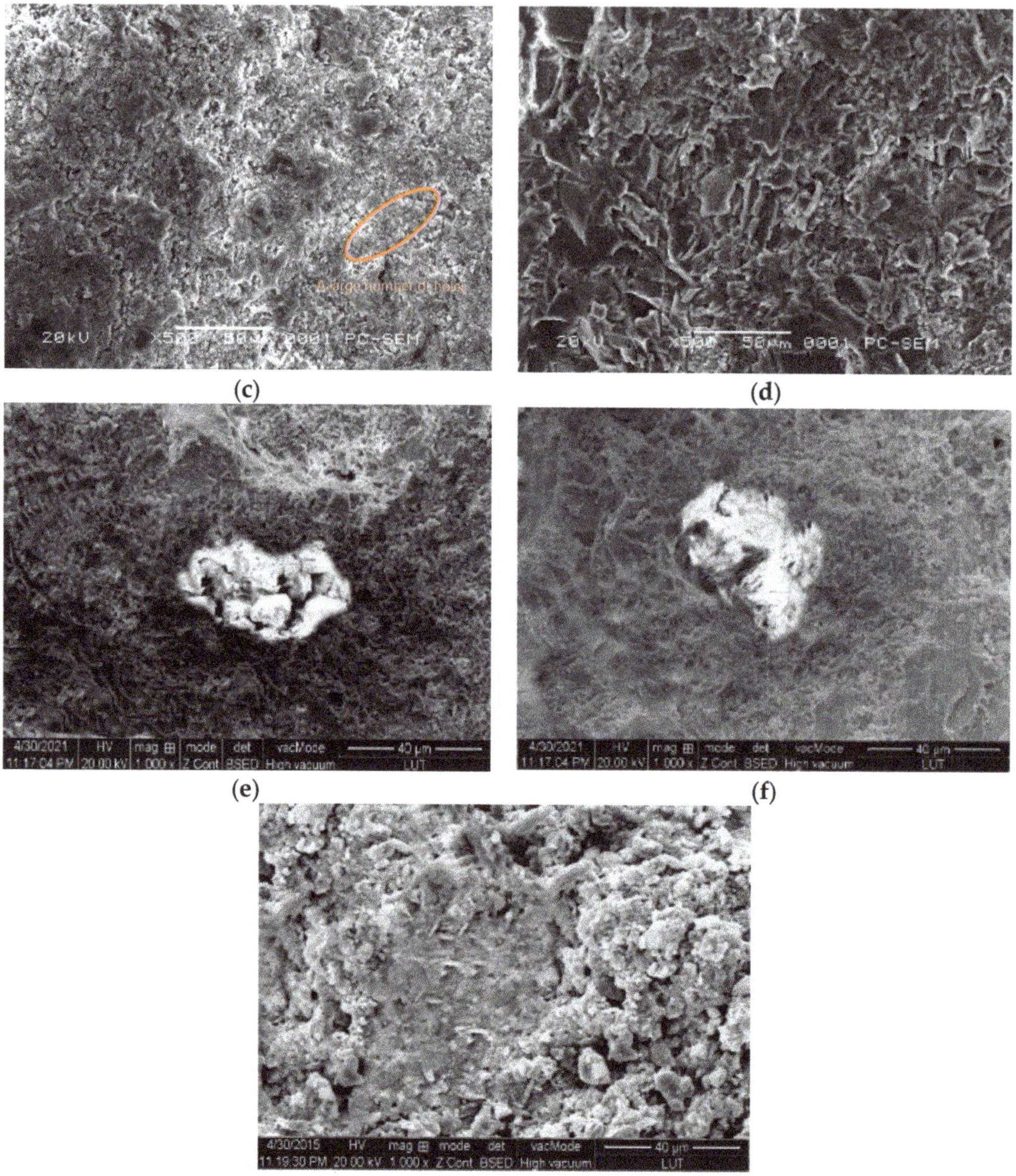

Figure 9. SEM photographs of rock after 60 F–T cycles for three treatments: (**a**) untreated samples (500×); (**b**) snowmelt (500×); (**c**) rain infiltration (500×); (**d**) normal F–T (500×); (**e**) corrosion of feldspar (1000×); (**f**) corrosion of calcite (1000×); and (**g**) increase in clay (1000×).

5. Discussion

The main visual weathering forms of the petroglyph were erosion (76.6%), peeling (12.7%), biological growth (9.5%) and man-made damage (1.1%). The weathering damage including granular disintegration and rounding/notching to make petroglyph lines is difficult to distinguish in natural light. The snowmelt, rainfall infiltration, and freeze–thaw processes are the critical causes of rock weathering in this area. The experiments found that the grains in the surface of the petroglyph corrode in snowmelt and rainfall infiltration form granular disintegration and notching [27]. In addition, the tests also detected that the strength of rock of petroglyph is easily affected by snowmelt, rainfall infiltration, and freeze–thaw possesses, which results in the production of cracks and peeling.

Mass loss, wave velocity and uniaxial compression tests were applied to study the processes in the rocks of petroglyph subjected to snowmelt, rainfall infiltration, and freeze–thaw processes. These tests show that the mass, wave velocity and compressive strength decrease with the number of F–T cycles. The associated parameters (determined by weathering index, wave velocity, stress-strain curve and compressive strength) of rocks in the snowmelt group were smaller than those of the other groups, and parameters for the rainwater infiltration group were smaller than those of the freeze–thaw group [28,29]. However, the rate of mass change shows the opposite pattern, i.e., the mass change is greatest in snowmelt group, which was confirmed by the SEM analysis. After the three treatments, the roughness of the rock surface between snow, rain and rock was obviously large, with the greatest roughness in the snowmelt group, followed by the rainwater infiltration group and freeze–thaw group. It turns out that snowmelt mainly affects the rock surface, making it rougher. Additionally, the mineral changes also provide clear evidence, i.e., the corrosion of calcite and feldspar and increase in clay are the main reasons for the increase in voids resulting a decrease in mechanical strength. The XRD analyses also indicate that calcite is highly susceptible to freeze–thaw action. The calcite content decreased significantly in the early stages of the tests, with the snowmelt group declining the fastest. However, the feldspar content decreased slowly in the all experiments, and was stable in freeze–thaw group for it may be associated with the type of ion and its content in snow and rain [19].

In these experiments, it is difficult to collect and store the snow and rain from the Helan Mountains and process many specimens for mechanical testing. For example, the specimens for the mass, velocity, SEM and XRD tests could be reused, but the specimens for testing mechanical strength were required in large quantities and were damaged by testing. As a consequence, the authors attempted to determine the relationships between the mass change rate, velocity change rate, and uniaxial compression between the snowmelt group and rainfall infiltration group with data from the freeze–thaw group. For each regression, the coefficients of fit and determination are shown in Figure 10. These graphs demonstrate that, in all situations, lineal and logarithmic regression curves were judged to be the most accurate representations of the relationships, which indicates that good linear and logarithmic correlations were obtained, especially among mass change rate, wave velocity change range and uniaxial compressive strength in group Y and group Z. The equations for the correlations are shown in Table 2 below.

The t test calculated with SPSS was performed among the produced equations to examine the validity of the regression equations used in this study (Equations (3)–(8)). The results show that there is a real correlation between the mass change, wave velocity change and the intensity of the snowmelt group and the rainfall infiltration group and the corresponding parameters of the freeze–thaw group, which can be used at least for the preliminary evaluation of the weathering process of snowmelt and rainfall infiltration.

Table 2. The analyses results by t test.

Equation Number	Regression Equation	Determination Coefficient (R2)	t Test	
			Calculated Value	Tabulated Value
(3)	$\Delta m_X = -0.0039 + 1.2767\Delta m_Z$	0.995	18.153	±1.35
(4)	$\Delta m_Y = -0.0127 + 1.896\Delta m_Z$	0.974	19.238	±1.53
(5)	$\Delta V\_X = 11.32 * EXP((V\_Z/17.55) - 11.32$	0.999	18.762	±1.25
(6)	$\Delta V\_Y = 12.81 * EXP(V\_Z/19.93) - 12.59$	0.999	18.531	±1.28
(7)	UCSX = 184.83 − 2.62E8lnUCSZ	0.997	18.319	±1.31
(8)	UCSY = 184.16 − 4.95E8lnUCSZ	0.999	18.256	±1.29

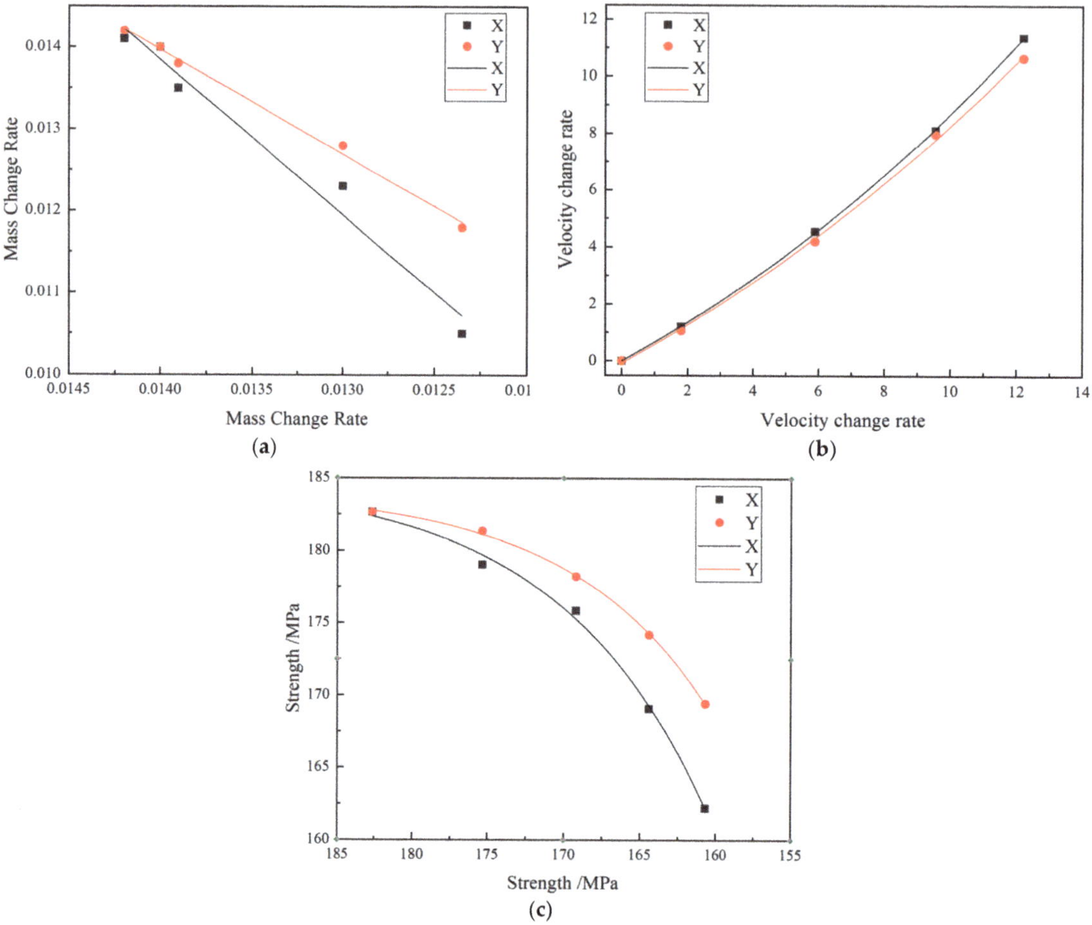

Figure 10. Correlation after X and Y with Z. (**a**) Correlation between the mass change rates. (**b**) Correlation between the velocity change rates. (**c**) Correlation between the uniaxial compressive strengths.

6. Conclusions

(1) The mass, wave velocity, stress–strain curve, compressive strength, mineral content, and microstructure were altered clearly in response to snowmelt and rainfall infiltration processes. Meanwhile, the variation in mass, particle size, porosity and pore width in the snowmelt group was greater than in the rainfall infiltration group, and the values for the rainfall infiltration group were greater than those in the normal freeze–thaw group. However, the wave velocity and uniaxial compression strength followed the reverse trend.

(2) The ordinary freeze–thaw treatment had a greater effect on the physical and mechanical properties of the petroglyph rocks than the snowmelt and rainwater treatments, as shown in the mass change, XRD and SEM tests. However, calcite and feldspar are prone to corrosion in the early stage of the snowmelt and rainfall infiltration processes. Thus, snowmelt and rainfall infiltration mainly affected the rock surface, making it rougher.

(3) Our study showed that the uniaxial compressive strength after normal freeze–thaw cycles had good accuracy for estimating the uniaxial compressive strength after snowmelt and rainfall infiltration, thus making it possible to assess the other two kinds of freeze–thaw processes. As a result, performing a freeze–thaw tests with snowmelt

or rainfall infiltration could be avoided, which are laborious, time-consuming and difficult to realize.

This study conducted preliminary research on the distribution and characteristics of damage to petroglyphs in the Helan Mountains. The deterioration of petroglyphs is a long process, which is undoubtedly a great challenge to scientific researchers. The protection of petroglyphs is difficult. Therefore, approaches to protect and reinforce the petroglyphs still need to be explored.

Author Contributions: Conceptualization, G.W. and K.C.; methodology, G.W.; validation G.W. and X.Y.; formal analysis, C.G.; investigation, G.W.; resources, G.W.; data curation, X.Y.; writing—original draft preparation, G.W.; writing—review and editing, G.W. and C.G.; visualization, X.Y.; supervision, K.C.; project administration, G.W.; funding acquisition, G.W. All authors have read and agreed to the published version of the manuscript.

Funding: This study was supported by National Natural Science Foundation of China (Grant No. 42107206), the Program for Changjiang Scholars and Innovative Research Team at the University of the Ministry of Education of China (No. 2017IRT17051), the National Natural Science Foundation of Gansu (Grant No. 22JR5RA292) and the "Hongliu Excellent Young" Talents Support Program of Lanzhou University of Technology.

Institutional Review Board Statement: Not applicable.

Informed Consent Statement: Not applicable.

Data Availability Statement: Not applicable.

Conflicts of Interest: The authors declare no conflict of interest.

References

1. Li, X. *Helan Mountain Rock Paintings Conservation Research Engineering Series: World Rock Paintings Appreciation*; Ningxia People's Publishing House: Ningxia, China, 2017; Volume 360.
2. Xu, F.; Yang, J.; Yang, Y. Study on the weatherability of hydraulic Lime as reinforcing material of Helan mouth's rock paintings. *Sci. Conserv. Archaeol.* **2016**, *28*, 31–39.
3. Zhang, W.; He, M.; Li, Y.; Cui, Z.; Wang, Z.; Yu, Y. Characteristics of quaternary glaciers in Helan mountains and their coupling with climate and tectonics. *Chin. Sci. Bull.* **2012**, *57*, 2390–2402.
4. Zhu, T.; Chen, J.; Huang, D.; Luo, Y.; Li, Y.; Xu, L. A DEM-based approach for modeling the damage of rock under freeze–thaw cycles. *Rock Mech. Rock Eng.* **2021**, *54*, 2843–2858. [CrossRef]
5. Zhu, C.; Xu, X.; Liu, W.; Xiong, F.; Lin, Y.; Cao, C.; Liu, X. Softening damage analysis of gypsum rock with water immersion time based on laboratory experiment. *IEEE Access* **2019**, *7*, 125575–125585. [CrossRef]
6. Huang, S.; Cai, Y.; Liu, Y.; Liu, G. Experimental and Theoretical Study on Frost Deformation and Damage of Red Sandstones with Different Water Contents. *Rock Mech. Rock Eng.* **2021**, *54*, 4163–4181. [CrossRef]
7. Chen, T.C.; Mori, N.; Suzuki, T.; Shoji, T.; Goto, T. Experimental study on crack development of rock specimens by freezing and thawing cycles. *Soils Found.* **2000**, *40*, 41–48. [CrossRef] [PubMed]
8. Chen, T.C.; Yeungb, M.R.; Mori, N. Effect of water saturation on deterioration of welded tuff due to freeze-thaw action. *Cold Reg. Sci. Technol.* **2004**, *38*, 127–136. [CrossRef]
9. Amoroso, G.G.; Fassina, V. Stone decay and conservation. In *Materials Science Monograph*; Elsevier: Amsterdam, The Netherlands, 1983; Volume 2.
10. Lu, Y.; Chen, X.; Tang, J.; Li, H.; Zhou, L.; Han, S.; Ge, Z.; Xia, B.; Shen, H.; Zhang, J. Relationship between pore structure and mechanical properties of shale on supercritical carbon dioxide saturation. *Energy* **2019**, *172*, 270–285. [CrossRef]
11. Kuila, U.; Prasad, M. Specific surface area and pore-size distribution in clays and shales. *Geophys. Prospect.* **2013**, *61*, 341–362. [CrossRef]
12. Nicholson, D.T.; Nicholson, F.H. Physical deterioration of sedimentary rocks subjected to experimental freeze–thaw weathering. *Earth Surf. Process. Landf. J. Br. Geomorphol. Res. Group* **2000**, *25*, 1295–1307. [CrossRef]
13. Tan, X.; Chen, W.; Yang, J.; Cao, J. Laboratory investigations on the mechanical properties degradation of granite under freeze–thaw cycles. *Cold Reg. Sci. Technol.* **2011**, *68*, 130–138. [CrossRef]
14. Zhang, J.; Miao, L.; Yang, Z. Research on rock degradation and deterioration mechanisms and mechanical characteristic under cyclic freezing-thawing. *Chin. J. Rock Mech. Eng.* **2008**, *27*, 1688.
15. Zhou, K.P.; Zhang, Y.M.; Li, J.L.; Xu, Y.J. Granite microstructure deterioration characteristic under condition of freezing-thawing based on NMR technology. *J. Cent. South Univ.* **2013**, *44*, 3384–3389.

16. Li, J.; Zhou, K.; Zhang, Y.; Xu, Y. Experiment study on physical characteristics in weathered granite under freezing-thawing cycles. *J. Cent. South Univ.* **2014**, *45*, 798–802.
17. Pu, T.; Chen, W.; Du, Y.; Li, W.; Su, N. Snowfall-related deterioration behavior of the Ming Great Wall in the eastern Qinghai-Tibet Plateau. *Nat. Hazards* **2016**, *84*, 1539–1550. [CrossRef]
18. Cui, K.; Wu, G.P.; Du, Y.M.; An, X.; Wang, Z. The coupling effects of freeze-thaw cycles and salinization due to snowfall on the rammed earth used in historical freeze-thaw cycles relics in northwest China. *Cold Reg. Sci. Technol.* **2019**, *160*, 288–299. [CrossRef]
19. Cui, K.; Liu, G.; Wu, G.; Zhu, P. Study on the characteristics and mechanism of freeze-thaw damage of rock carrier in Helan mouths rock paintings under different conditions. *Chin. J. Rock Mech. Eng.* **2019**, *38*, 1797–1808.
20. Wang, X.D.; Zhang, B.; Pei, Q.Q.; Guo, Q.L.; Chen, W.W.; Li, F.J. Experimental studies on sacrificial layer in conservation of earthen sites. *J. Cult. Herit.* **2020**, *41*, 74–83. [CrossRef]
21. Stewart, I.T. Changes in snowpack and snowmelt runoff for key mountain regions. *Hydrol. Process.* **2009**, *23*, 78–94. [CrossRef]
22. Chen, W.W.; Wei, D.C.; Lei, H.; Li, Y. Experimental study on strength deterioration characteristics of earthen sites covered by snow. *J. Lanzhou Univ.* **2019**, *55*, 655–660. (In Chinese)
23. Xu, F. *Rescue Protection Scheme of Rock Paintings in Helan Mountain*; Nanjing Museum: Nanjing, China, 2008; Volume 230. (In Chinese)
24. Wang, P. On the protection of rock paintings in Helan mountain from the cause of diseases. *Restor. Res. Cult. Relics* **2014**, *00*, 587–596. (In Chinese)
25. ICOMOS-ISCS. Illustrated glossary on stone deterioration patterns. In *ICOMOS-ISCS: Glossaire Illustré Sur les Formes d'altération de la Pierre*; ICOMOS: Paris, France, 2001; Volume 15, 78p, ISBN 978-2-918086-00-0.
26. *GBT50266-2013*; Standard for Test Method of Engineering Rock Mass. Administration of Quality Supervision, Inspection and Quarantine of People's Republic of China: Beijing, China, 2013. (In Chinese)
27. Mol, L. Investigations into the relationship between changes in internal moisture regimes and rock surface deterioration in cavernous sandstone features. *Earth Surf. Process. Landf.* **2014**, *39*, 914–927. [CrossRef]
28. Li, J.; Kaunda, R.B.; Zhou, K. Experimental investigations on the effects of ambient freeze-thaw cycling on dynamic properties and rock pore structure deterioration of sandstone. *Cold Reg. Sci. Technol.* **2018**, *154*, 133–141. [CrossRef]
29. Cui, K.; Wu, G.; Wang, X.; Chen, W. Behaviour of slate following freeze–thaw and dry–wet weathering processes. *Q. J. Eng. Geol. Hydrogeol.* **2017**, *50*, 117–125. [CrossRef]

coatings

MDPI

Article

Scientific Research on a Gold- and Silver-Inlaid Bronze Zun from the Han Dynasty

Dan Liu , Xiaolong Tian, Dong Zhang *, Xianjing Zhou, Nana Li and Yajun Zhao

Cultural Relics Conservation and Restoration Center, Gansu Provincial Museum, Lanzhou 730050, China; ld1153@sina.com (D.L.)
* Correspondence: yuren1515@163.com; Tel.: +86-13669333418

Abstract: The bronze Zun was one of the more prevalent high-class wine containers of the Han dynasty, representing the highest level of decoration in bronze at the time. However, little has been reported about its technical characteristics and scientific value. In this paper, the samples were selected for analysis based on scientific analysis, following the principle of "minimal intervention", and a bronze Zun from the Han dynasty in the Gansu Provincial Museum collection was studied using ultra-deep field microscopy, X-ray flaw detection, X-ray fluorescence spectrometry (XRF), scanning electron microscopy (SEM), and energy spectrometry (EDS). The results show that the gold and silver decoration on the bronze is inlaid rather than gilt. Secondly, the body and lid of the vessel are molded in one shot, with the bird-head-shaped and animal-foot-shaped components cast separately and then attached to the lid and body. Thirdly, the corrosion of the bronze Zun is characterized by the copper matrix being corroded first and most severely, followed by the silver and, finally, the gold. The high purity of the gold wire embedded in this bronze Zun, the fine width of gold wire (154–190 μm), and the magnificent decoration show the excellent processing technology level of the precious metal and the high aesthetic level of ancient man during the Han dynasty. The results of the analysis of this bronze Zun can provide an essential reference for research on bronze vessels of the same type, the techniques of gold and silver misalignment, and the development of the history of bronze manufacture and technology during the Han dynasty.

Keywords: Han dynasty; bronze Zun; gold and silver inlaying; scientific research

1. Introduction

The Zun (bronze vessel) was a practical everyday wine vessel that appeared during the Warring States period (476–221 BC) and flourished during the Han dynasty (202 BC–8 AD). It came in two shapes similar to basins or barrels, mostly with three feet or a ring foot [1–3]. The bronze Zun was a more expensive, high-class wine container during the Han dynasty and was one of the most prominent wine containers. In ancient China, there was a tradition of using wine to entertain guests, express personal feelings, and even treat diseases. China's wine culture began almost along with the splendid civilization of 5000 years [4]. Because of this penchant for wine, the imperial aristocracy was very particular about wine vessels under the social conditions of the time. As a result, the popular Zun during the Han dynasty was particularly prestigious at the time, representing the Han bronze wares' highest level of decoration.

The bronze Zun of the Han dynasty enjoyed unprecedented development and reached a high level of decoration based on the techniques used to make it. Standard decorative techniques of bronze ware include inlay [5,6], gilding [7], gold inlaying [8], gold wrapping [9], burin engraving [10], and so on. However, because of the paucity of surviving ancient texts, there is still further research required into gold inlaying [11]. From the excavated cultural relics, we can understand that the ancient misprinted gold and silver process includes two forms of gold inlaying and gilding [12]. Inlay uses gold wire, gold plates, silver wire, silver plates, and other decorative objects on cast bronze inlaid with various ornaments or inscriptions. The gilding system, sometimes called the "gold coating method", uses gold amalgam on the surface of the bronze, which is repeatedly coated with gold; after the mercury evaporates, the gold remains on the object's surface [13]. In connection with this, gilded objects decorated with gold may also have gold inlaying, and those gilded with silver may also have silver inlaying. The process of gold and silver inlaying is more complicated than gilding and requires large amounts of gold and silver, so gold- and silver-inlaid bronze is precious; studying such objects is of significant value to improve our understanding of their craftmanship.

There are some studies on the gold and silver inlaying and gilding processes on bronze at home and abroad. Based on the available research, the basic steps of the gilding method used in different regions are approximately the same [14–17]. A study by Oddy et al. on the composition of decoration on gold vessels, silver vessels, and bronze vessels from the ancient and medieval periods found that the statue of the Roman emperor Niello (1st century AD) was made of bronze. At the same time, the object's surface was inlaid with silver and had a shallow groove with a black coating, known as "Niello". The technical analysis carried out by Konstantinidi-Syvridi E. and others, based on archaeological and experimental data, revealed that a decorative technique using gold inlay on the surface of luxury weapons, called gold embroidery, was prevalent in the early Mycenaean period [18]. This metalworking technique involved the dense placement of tiny gold wires or particles of gold (also known as gold bars) next to each other and, finally, engraving them into a decorative pattern. Although this is a gold inlaying technique, it differs from what we have described as gold inlay. It is worth noting that the dimensions of the bars used in this decoration technique range from 0.3 to 8 mm in length, 0.4 to 1.5 mm in width, and 0.3 to 0.6 mm in thickness, respectively, a clear difference from Chinese Han dynasty metalworking techniques. Yang et al. studied the gilding process of a Warring States bronze chime from the Sichuan Museum collection. They speculated that the gold wire of chime bells was decorated with groove-gilding technology [19]. Liu et al. carried out the conservation and restoration of a gold-inlaid silver belt hook excavated during the Warring States period based on modern instrumental analysis methods [20]. However, the available research findings have yet to explore the depth of the gold and silver inlaying process on bronze Zun.

The conservation of cultural relics is not only about treating the damage but also about extracting as much information as possible about the relics and digging deeper into their artistic, historical, and scientific values [21]. Based on the above, we have undertaken a scientific analysis of a Han dynasty gold- and silver-inlaid bronze Zun from the Gansu Provincial Museum collection to provide a reference for research on the gold inlaying process of similar types of bronzes and the scientific and technological history of bronze production during the Han dynasty.

2. Materials and Methods

2.1. Materials

Figure 1 shows the appearance of the gold- and silver-inlaid bronze Zun. It is cylindrical, with a lid, a flat base, and three animal-shaped feet. The overall weight of the object is 2.1 kg, of which the lid weighs 0.54 kg and the body weighs 1.56 kg. The overall height is 20 cm, of which the body is 15 cm high and the lid is 5 cm high. The outer diameter of the body is 16.6 cm, and the inner diameter of the mouth rim is 14.2 cm. The top of the lid is surmounted by a looped handle, with three buttons outside the loop, and the central area is carved in relief with a persimmon peduncle, with symmetrical classes with a ring handle on the belly. Neither the lid nor the base of the vessel is decorated or inscribed. In addition, the object's surface is covered with a green and black patina due to its burial and preservation environment, which has caused some of the decorative motifs on the vessel's surface to be covered. It is tentatively determined that the wires removed from the body of the bronze Zun and lid are gold and silver.

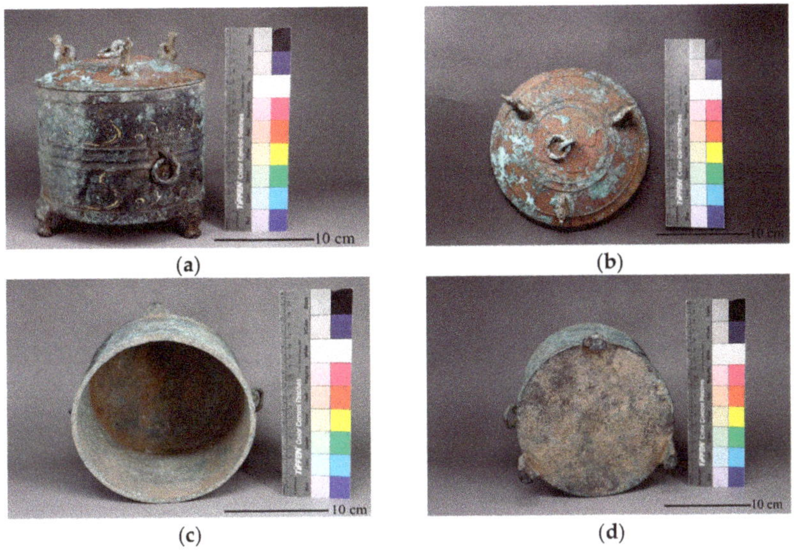

Figure 1. Appearance of the bronze Zun. (**a**) is overall appearance; (**b**) is lid appearance. (**c**) is internal body. (**d**) is bottom of the body.

2.2. Methods

The analysis and sampling for this study were carried out using non-destructive or minimally invasive methods. Therefore, a digital VHX–6000 microscope (KEYENCE, Osaka, Japan) was used for microscopic observation. The bronze vessel was photographed fluoroscopically by Smart Evo 300D X-ray flaw detection (YXLON, Hamburg, Germany) to observe its internal structure and the information on the surface decoration covered by rust without special treatment of the samples. Non-destructive testing of the alloy composition of the bronze Zun was performed using X-ray fluorescence spectroscopy (Thermofisher, Waltham, MA, USA). The body was examined in the normal metal mode, and the gold and silver wires were examined in the precious metal mode with a test time of 60 s. The microscopic morphology of the samples was observed by scanning electron microscopy (SEM) (JSM–6610LV, JEOL, Tokyo, Japan), and the matrix and inclusions of the samples were analyzed by X-ray spectrometry (INCA X-ACT 250, Oxford Instruments, High Wycombe, UK) for their micro-zone chemical composition.

3. Results and Discussion

3.1. Ultra-Deep Field Microscope Analysis

Figure 2a shows the gold wire coming off the edge of the lid of the Zun and the exposed grooves, which would have been formed by a chisel. It can be seen that the burin grooves are wide on the outside and narrow on the inside, which is different from the usual grooves in decorated gold, which are wide on the inside and narrow on the outside. The different widths and depths of the chisel grooves also illustrate the random nature of the ancient artisan's chiseling operation. The gold filigree is essentially flaking off as a whole with little residual trace, and it can be tentatively inferred that this would not have been formed by the gilding method of gold amalgam. In addition, measurements show that the width of the gold wire varies between 154 and 190 μm (Figure 2b), which is much smaller than the size of the gold used in the aforementioned Mycenaean weapons inlaid with gold, indicating the high level of metalworking technology of the Han dynasty [18]. Figure 2c shows a detailed view of the articulation of the gold wire, which can be seen at the junction where the wire is overlapped, with apparent traces of jointing work, which may be a remedial measure for the artisan's underestimation of the length of the wire when it was inlaid and which also indicates the use of smashing and pounding in the process of inlaying the gold wire. The uneven edges of the gold can also be seen, with traces of cuttings. As can be observed in Figure 2d, there are clear, congruent polishing scratches on the gold filigree's surface, which the whetstone grinding created. In addition, the edge of the chiseled groove at the location of the gold and silver is tilted inward, presumed to be caused by hammering to make the malleable gold and silver pieces more closely connected to the chiseled groove.

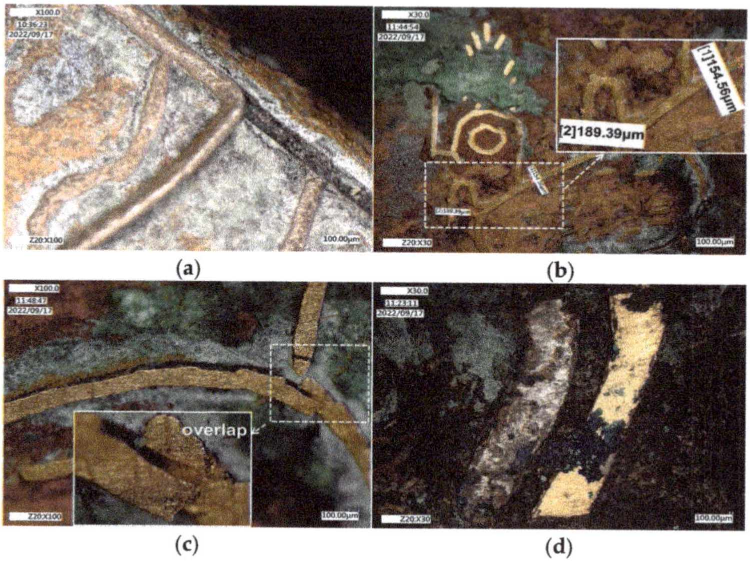

Figure 2. Ultra-deep field microscope photos of the bronze Zun: (**a**) the chisel groove; (**b**) the gold wires ([1] and [2] refer to the width of the gold wire measured at 2 different locations.); (**c**) the junction of gold wires; and (**d**) the surface grinding marks of the gold and silver pieces.

3.2. X-ray Flaw Detection Analysis

In order to gain further insight into the overall manufacturing craft of the bronze Zun, it was tested and analyzed using X-ray flaw detection imaging inspection techniques, the results of which are shown in Figure 3. The delicate ornamental motifs and decorations on the lid and body can be seen on the X-ray flaw detection graphs. The surface of the lid is decorated with bird and animal motifs as the prominent motifs, and it is decorated with a combination of cut diamond-shaped pattern pieces and gold thread to form auxiliary motifs. The main decoration and auxiliary decoration patterns on the surface of the bronze Zun are clearly layered, which is a typical pattern of decoration in the Han dynasties [22]. The bronze Zun is decorated with a bow-string pattern, a diamond-shaped pattern, and wavy patterns around it, with three animal-head-shaped components at the base to provide support (Figure 3a,b). Due to the photographic limitations of X-ray flaw detection, it is difficult to distinguish the patterns of the front and back sides of the bronze Zun with cavity structure, and the decorative motifs around the body are superimposed onto the same plane while taking an X-ray film. Despite this limitation, fine surface decoration and motifs can be observed (Figure 3b). The "SS" decoration on the body is also visible in Figure 3b, consisting of cut gold and silver filigree. The shades of color of the "SS" motifs in the X-rays vary, and it is speculated that the darker and lighter motifs are silver and gold, as the higher the atomic mass and the higher the density, the lower the blackness of its position [23]. This is consistent with the results of the exterior photographs of the objects (Figure 1a) and the super-field microscopic analysis (Figure 2d), all of which demonstrate the fine line engraving and delicate decorative techniques that were part of the superb handwork of ancient bronze decoration. However, varying degrees of detachment and loss of gold and silver filigree and flakes can also be observed.

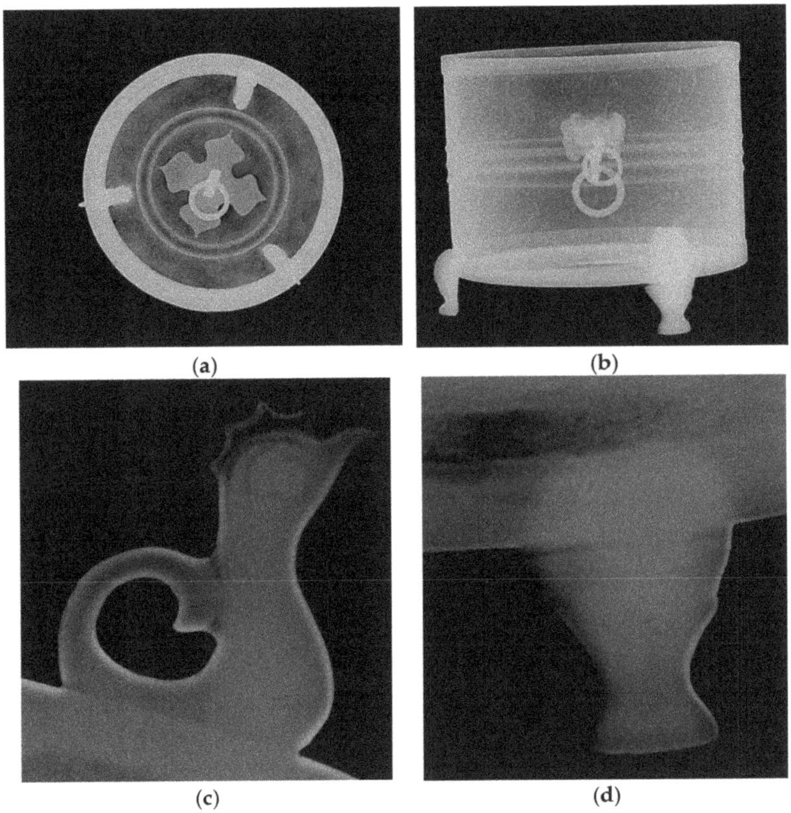

Figure 3. X-ray flaw detection graphs of the bronze Zun: (**a**) the lid of the bronze Zun; (**b**) the body of the bronze Zun; (**c**) the bird-head-shaped component on the lid; and (**d**) the animal-foot-shaped component on the body of the bronze Zun.

From Figure 3a, the lid has a relatively dense texture, indicating a relatively homogeneous copper liquid during the casting process and a stabilized subsequent cooling process. Four shading spots on the lid are presumed to be possible shrinkage holes created during the casting process [24]. This is because, during the solidification and shrinkage of the coppery liquid, the thinner-walled locations would have solidified first, thus blocking the passage for the flow of the coppery liquid. The subsequently solidified locations would not have been able to replenish the coppery liquid as they continued to cool and shrink, resulting in air pockets and shrinkage holes.

In addition, the difference in the degree of X-ray absorption between the bird-head-shaped components at the top of the lid and the body of the lid and between the animal-foot-shaped components and the main body of the vessel should be due to the different densities of the materials (Figure 3c,d) [25]. Combined with the appearance photos of the lid and the body (Figure 1), no sign of a match was observed, and it is presumed that molding was used in divided model-making. Therefore, it seems as though the main body of the bronze Zun was cast once, and the three bird-head-shaped and three animal feet of the lid were cast separately and then connected.

3.3. X-ray Fluorescence Spectroscopy Analysis

The lid base, the body, the bird-head-shaped components, and the animal-foot-shaped components of the bronze Zun were selected for surface alloy composition testing. Table 1 presents the percentage content of the elements and the respective standard deviations (RSDs) obtained in both bronze samples. The main elements in this bronze Zun are copper, tin, and lead, with 73.28%, 21.03%, and 3.05% in the base of the lid and 77.00%, 17.66%, and 1.91% in the base of the body, indicating that the bronze is a ternary Cu-Sn-Pb alloy and a high-tin bronze [26]. The copper, tin, and lead contents of the bird-head-shaped lid and the foot of the animal on the body are 66.55%, 18.12%, 1.91%, 80.63%, 15.89%, and 0.64%, respectively. The difference between the elemental content of the base of the lid and that of the body of the bronze Zun confirms, on the one hand, the findings of the X-ray analysis that the bronze Zun and the lid would have been cast in one piece and that the bird-head-shaped and animal-foot-shaped components were cast separately and then attached to the lid and body. On the other hand, the rust on the surface of this bronze may have interfered with the XRF results to some extent.

Table 1. XRF results of each part of the bronze Zun (wt %).

No.	Test Content	Fe	Cu	Pb	Sn
1	Lid body	0.07	73.28	3.05	21.03
	(RSD%)	(21.02)	(2.73)	(11.52)	(9.57)
2	Bird-head-shaped components	0.51	66.55	1.92	18.12
	(RSD%)	(17.93)	(3.97)	(13.66)	(11.73)
3	Vessel body	2.90	77.00	1.91	17.66
	(RSD%)	(11.25)	(3.82)	(13.98)	(12.11)
4	Animal-foot-shaped components	2.41	80.63	0.64	15.89
	(RSD%)	(10.49)	(2.35)	(18.01)	(12.56)

3.4. SEM–EDS Analysis

Figure 4 shows the SEM images of the gold and silver wires. SEM micro-zone analysis reveals traces of shear processing on the surface of the gold wire, a process that was used to better embed it into the surface of the bronze (Figure 4a,b), with impurities adhering to the surface. After initial cleaning of the cross-section of the gold wire, it can be seen that the entire profile is relatively homogeneous, with tiny particles of gold (Figure 4c). Meanwhile, an uneven surface of the silver wire can be observed (Figure 4d), and particle attachment is also evident from the center of Figure 4e. The profile of the silver wire shows a more uniform distribution, similar to that of the gold wire (Figure 4f). This is because both gold and silver have the same face-centered cubic crystal structure, as well as similar cell parameters and chemical properties [27].

The analysis results of the surface and profile elements of the gold and silver wires and their relative standard deviations are shown in Table 2. The surface of the gold wire contains 67.48%, 7.68%, 3.61%, 4.42%, 1.40%, 14.86%, and 0.91% of gold, silver, copper, tin, chlorine, oxygen, and aluminum, respectively. The gold wire profile contains 92.50% gold and 7.50% silver, which can be inferred to be pure gold. This is because gold in its natural state, whether obtained from a placer or vein gold mine, usually contains considerable amounts of silver (usually 5.00% to 45.00% by weight) and other major associated impurities of gold [28]. Some scholars believe that the small amount of silver detected in the gold is more likely to represent a natural impurity in the gold than an artificially prepared alloy [29]. However, to conclude that the gold used for decoration in this bronze is natural gold smelted or synthesized through the later artificial addition of silver, more bronze or gold and silver vessels of the same type need to be analyzed.

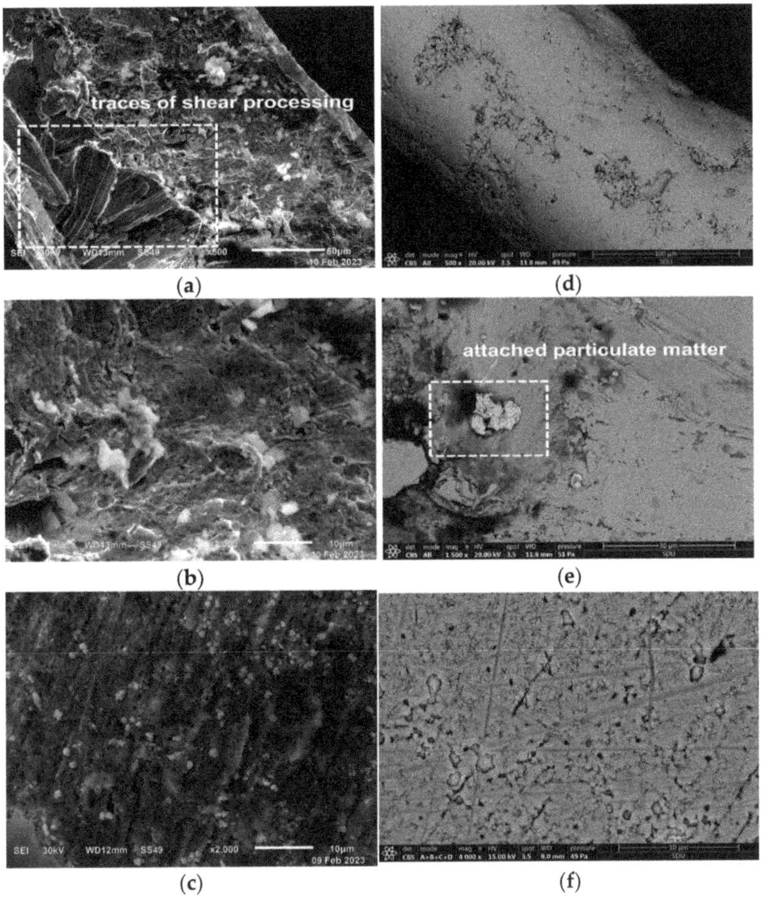

Figure 4. SEM photos of the gold and silver wires: (**a**,**b**) the surface of the gold wire; (**c**) the gold wire profile; (**d**,**e**) the surface of the silver wire; and (**f**) the silver wire profile.

Table 2. SEM–EDS results of the composition analysis of the gold and silver wires.

Test Content	wt%								
	Au	Ag	Cu	Sn	Cl	S	Si	Al	O
the surface of the gold wire	67.48	7.68	3.61	4.42	1.04	-	-	0.91	14.86
(RSD%)	(9.52)	(10.76)	(11.43)	(11.21)	(13.20)			(15.76)	(9.67)
the gold wire profile	92.50	7.50	-	-	-	-	-	-	-
(RSD%)	(5.33)	(9.47)							
particulate matter on the surface of the silver wire	-	99.60	0.40	-	-	-	-	-	-
(RSD%)		(3.22)	(16.87)						
particulate matter on the silver wire	-	11.00	71.00	16.00	1.40	0.30	0.20	0.10	-
(RSD%)		(11.73)	(7.59)	(10.56)	(13.61)	(17.24)	(18.57)	(19.0)	
the silver wire profile	-	78.30	-	-	21.7	-	-	-	-
(RSD%)		(8.12)			(10.71)				

Furthermore, the Cu and Sn elements contained on the surface of the gold wire would have come from the body of the bronze Zun. The traces of Cl and Ag elements detected on the surface of the gold wire could have been caused by the corrosion of the silver contained in the gold wire, which is probably in the form of silver–chlorine species [30,31]. On the one hand, the silver content on the surface of the silver wire is 99.60% silver and 0.40% copper. The silver, copper, chloride, and tin contents of the particulate matter on the surface of the silver wire are 11.00%, 71.00%, 1.40%, and 16.00%, respectively. The detected copper element would result from the silver wire being embedded in the bronze and prolonged contact with the bronze matrix of the vessel body. On the other hand, the main constituent elements of the energy spectrum are O, Ag, Cl, and Cu, and it is presumed that the particulate matter is mainly cuprous oxide and silver chloride. The silver wire profile of 78.30% silver and 21.70% chloride suggests that bronze is more severely corroded.

By comparing and analyzing the elemental composition and content of the profiles of gold and silver wire, it was found that the content of gold was higher than that of silver, and the silver wire was corroded a little more seriously. Combined with Figures 1 and 2, the copper matrix's corrosion was more severe than that of the silver wire, with more and thicker corrosion at the gold and silver inlay decoration. Because the gold- and silver-inlaid bronze Zun mainly contains Au, Ag, and Cu, three different metals, these three metals' electrode potentials are different ($E_{Au} > E_{Ag} > E_{Cu}$). According to electrochemical principles, it is known that when there is an electrolyte in a buried or preserved environment, the copper matrix is corroded first, followed by silver, and finally gold. Suppose the patina is treated directly with chemical cleaning agents on the one hand. In that case, this will result in the surface of the bronze being slightly lower than where the gold and silver are decorated, affecting the artifact's appearance. On the other hand, the cleaning solution may even enter between the gold and silver filigree and the bronze matrix, dissolving the patina therein and possibly causing the decoration to loosen or fall off. If the mechanical method removes the rust, it may leave scratches on the surface of the gold and silver decoration due to its soft texture [32]. During the conservation and restoration of the relics at a later stage, the pros and cons of these two methods should be weighed, and a reasonable conservation program should be developed.

In addition, Ag is susceptible to corrosion into AgCl, and this corrosion product can migrate to all parts of the object and even contaminate the surface of the artifact [33]. The absence of mercury in the SEM–EDS results leads us to conclude that the gold wire in this bronze Zun was not formed by the fire-gilding method, which supports our discussion in Section 3.1.

4. Conclusions

This study is based on morphological and chemical elemental analyses of a bronze Zun from the Han dynasty in the Gansu Provincial Museum. Through comparative analysis of the test results, the main conclusions were as follows:

(1) The gold and silver inlay decorations on this bronze Zun used techniques such as chiseling grooves, inlaying, and polishing rather than gilding;
(2) The body and lid would be cast in one piece, with the bird-head-shaped and animal-foot-shaped components cast separately and then attached to the lid and body;
(3) The bronze Zun, being a gold- and silver-inlaid bronze vessel, is characterized by the corrosion of the copper matrix first and most severely, followed by the silver, and then the gold;
(4) The high purity of the gold wire embedded in this bronze Zun, the fine width of gold wire (154–190 μm), the large amount of gold and silver used throughout, and the magnificent decoration reflect the abundance of gold minerals during the Han dynasty and the widespread dissemination and development of gold processing techniques during this period. At the same time, the technical and aesthetic level of precious metal processing is high. However, to determine whether the gold used in the decoration of this bronze is natural gold smelted or synthesized by the later

addition of silver, we believe that more bronze or gold and silver vessels of the same type need to be analyzed.

The analysis of this gold- and silver-inlaid bronze Zun is of great importance to studying human technology and lifestyle during the Han dynasty, as well as providing an essential reference for the conservation and restoration of similar bronze objects and the study of technology.

Author Contributions: Conceptualization, D.L. and X.T.; methodology, N.L. and X.Z.; investigation, Y.Z.; writing—original draft preparation, D.L.; writing—review and editing, D.L. and D.Z. All authors have read and agreed to the published version of the manuscript.

Funding: This research was funded by the Natural Science Foundation of Gansu Provincial Science and Technology Department (22JR5RA605) and the Gansu Provincial Museum collection metal cultural relics protection and restoration project (22-5-13-6200-412).

Institutional Review Board Statement: Not applicable.

Informed Consent Statement: Not applicable.

Data Availability Statement: Not applicable.

Acknowledgments: We are grateful to Qinglin Ma (Shandong University) for his guidance and Yunpeng Wang (Shandong University) for his assistance on the SEM–EDS analysis.

Conflicts of Interest: The authors declare no conflict of interest.

References

1. Jessica, R. Chu Influences on the Development of Han bronze vessels. *Arts Asiat.* **1989**, *44*, 84–99.
2. Sun, Y.-K.; Wu, I.-W.; Lin, R.-T. Transforming "Ritual Cultural Features" into "Modern Product Forms": A Case Study of Ancient Chinese Ritual Vessels. *Religions* **2022**, *13*, 517. [CrossRef]
3. Poo, M.-C. The Use and Abuse of Wine in Ancient China. *J. Econ. Soc. Hist. Orient* **1999**, *42*, 123–151. [CrossRef]
4. Wang, W.-F.; Wei, J.; Xu, H.; Zhang, Y.-D.; Chen, H. Relevance of Ancient Chinese Wine Ware Representation Design and Cultural Characteristics Based on Machine Learning and Semiotic Theory. *Wirel. Commun. Mob. Comput.* **2022**, *2022*, 2035662. [CrossRef]
5. Marianne, M.; Mauro, B.; Jarno, B.; Marco, F.; Martin, F.; Giorgia, G.; Martino, N.; Judith, U. Multidisciplinary analyses on the 11th-12th century bronze doors of San Marco, Venice. *PLoS ONE* **2023**, *18*, e0288094. [CrossRef]
6. Luo, W.-G.; Li, T.; Wang, C.-S.; Huang, F.-C. Discovery of Beeswax as binding agent on a 6th-century BC Chinese Turquoise-inlaid Bronze sword. *J. Archaeol. Sci.* **2012**, *39*, 1227–1237. [CrossRef]
7. Shao, Y.-B.; Lu, X.; Fu, W.-B.; Jiang, F.-R.; Yang, J.-C.; Gai, Z.-Y.; Dong, L.-M. Technical characteristics and coating formation mechanism of gilded silver products unearthed from the Consort Tomb of Emperor Shengzong of the Liao dynasty. *Archaeol. Anthropol. Sci.* **2023**, *15*, 28. [CrossRef]
8. Luo, W.-G.; Song, G.-D.; Hu, Y.-Q.; Chen, D. Tentative determination of a special bronze material by multiple technological test on a xuan-liu dagger-axe from the Xujialing Site, the Eastern Zhou period, Henan Province, China. *J. Cult. Herit.* **2020**, *46*, 304–312. [CrossRef]
9. Huang, M.; Wu, X.-T.; Chen, X.-Z.; Tao, L.; Wu, X.-H.; Shi, M.; Li, F.; Ritchey, M.-M.; Huang, F.; Jin, Z.-Y. Wuchuan bronzes and cinnabar mining immigrants during the Qin and Han Dynasties-new perspectives from typological and lead isotope analysis. *Archaeol. Anthrop. Sci.* **2021**, *13*, 198. [CrossRef]
10. Bagley, R.-W. Shang Ritual Bronzes: Casting Technique and Vessel Design. *Arch. Asian Art.* **1990**, *4*, 6–20.
11. Liu, Y.; Yang, J.-C.; Panpan, T. Some new thoughts about the technologies of "Cuojinyin". *Sci. Conserv. Archaeol.* **2019**, *31*, 75–86. [CrossRef]
12. Chen, A.-D. A brief talk on the East Hanbo Mountain cover god beast pattern copper bottle. *Silk Road* **2013**, *4*, 63–64. [CrossRef]
13. Yao, Z. Reunderstanding of ancient gold inlay and gilding techniques. *Huaxia Archaeol.* **2019**, *5*, 113–119. [CrossRef]
14. Masi, G.; Chiavari, C.; Avila, J.; Esvan, J.; Raffo, S.; Bignozzi, M.C.; Asensio, M.C.; Robbiola, L.; Martini, C. Corrosion investigation of fire-gilded bronze involving high surface resolution spectroscopic imaging. *Appl. Surf. Sci.* **2016**, *366*, 317–327. [CrossRef]
15. Tan, P.-P.; Yang, J.-C.; Ren, X.-L. Technical features of a ninth-century silver vessel of southern China uncovered from Famen Monastery, Shaanxi province. *Herit. Sci.* **2021**, *9*, 55–66. [CrossRef]
16. Brocchieri, J.; Scialla, E.; Manzone, A.; Graziano, G.O.; D'Onofrio, A.; Sabbarese, C. An analytical characterization of different gilding techniques on artworks from the Royal Palace (Caserta, Italy). *J. Cult. Herit.* **2022**, *57*, 213–225. [CrossRef]
17. Oddy, A.; Bimson, M.; Niece, S.L. The composition of niello decoration on gold, silver and bronze in the antique and mediaeval periods. *Stud. Conserv.* **1983**, *28*, 29–35. [CrossRef]
18. Eleni, K.; Nikolas, P.; Akis, G.; Maria, K. Gold Embroidery: A Sophisticated Technique for Early Mycenaean Swords and Daggers. *Archäologischer Anz.* **2022**, *2*, 1–127. [CrossRef]

19. Yang, J.; Wei, Q.; Zhang, Z.-J.; Cao, Y.-Y.; Wei, G.; Shao, Y.-B. Scientific research on gold-decoration techniques of Warring States bronze chime bells in Sichuan Museum. *Sci. Conserv. Archaeol.* **2022**, *34*, 69–79. [CrossRef]
20. Liu, X.-B.; Jiang, L.-M.; Zhou, X.; Li, P.; Xiao, L. Protection and restoration of gold and silver inlaid belt hook unearthed from Feihu Village, Pujiang County, Chengdu. *Sci. Conserv. Archaeol.* **2021**, *33*, 75–81. [CrossRef]
21. Li, J.; Li, K.; Zhao, F.; Feng, X.; Yu, J.; Li, Y.; Chao, X.; Wang, J.; Mai, B.; Cao, J. Three-Dimensional Laser Scanning Technology Assisted Investigation and Extraction of Human Bone Information in Archaeological Sites at Shenna Ruins, China. *Coatings* **2022**, *12*, 1507. [CrossRef]
22. Wu, S. *The Complete Collection of Chinese Patterns (Warring States·Qin·Han Scrolls)*; Shandong Fine Arts Publishing House: Shandong, China, 2009; pp. 3–40.
23. Wang, Q.-Y. Applications of X-radiography to the conservation and technical study of antiquities. *Sci. Conserv. Archaeol.* **2022**, *34*, 10–16. [CrossRef]
24. Basso, E.; Pozzi, F.; Day, J.; Borsch, L. Unmasking a wild man: Scientifc analysis of Bertoldo di Giovanni's Shield Bearer in The Frick Collection. *Herit. Sci.* **2020**, *8*, 109–122. [CrossRef]
25. Guo, R.; Feng, J.; Xiang, J.-K.; Dang, X.-J.; Zhang, X.-L.; Wang, Z.-L. Investigation on the casting process of bronzes in noble tombs of late Western Zhou Dynasty in Chang'an District, Xi'an. *Hua Xia Archeaol.* **2022**, *6*, 107–112. [CrossRef]
26. Omid, O.; Zeinab, K.; Zoya, K. The Metallic Sounds: A Microanalytical Study on the Production of Armenian Church Bells from Iran. *Microsc. Microanal.* **2023**, *29*, 1298–1306. [CrossRef]
27. Yang, X.-L.; Wang, H.-T. Study on the production technology of the unearthed parts of Danglu in the tomb of Liu He, the former emperor of the Western Han Dynasty. *Cult. Relics South. China* **2017**, *1*, 98–104. [CrossRef]
28. Scott, D.-A. The deterioration of gold alloys and some aspects of their conservation. *Stud. Conserv.* **1984**, *28*, 194–203. [CrossRef]
29. Photos, E.; Jones, R.-E.; Papadopoulos, T.H. The black inlay decoration on a Mycenaean bronze dagger. *Archaeometry* **1994**, *36*, 267–275. [CrossRef]
30. Scott, D.-A. *Metallography and Microstructure of Ancient and Historic Metals*; Oxford University Press: Cary, NC, USA, 1991; pp. 1–176.
31. Ingo, G.-M.; Balbi, S.; de Caro, T.; Fragalà, I.; Angelini, E.; Bultrini, G. Combined use of SEM-EDS, OM and XRD for the characterization of corrosion products grown on silver roman coins. *Appl. Phys. A* **2006**, *83*, 493–497. [CrossRef]
32. Shao, A.-D. Consideration on the protection of gold and silver inlaied bronze artifacts unearthed by archeologists. *Chin. Cult. Herit. Sci. Res.* **2010**, *1*, 58–60. [CrossRef]
33. Wu, H.-T.; Zhou, S.-L. Scientific analysis of gold-plated silver objects unearthed from the Tang dynasty Tubo tomb in Dulan County, Qinghai Province. *Sci. Conserv. Archaeol.* **2014**, *26*, 69–75. [CrossRef]

coatings

MDPI

Article

Investigation of Whitening Mechanism on Cultural Relic Surfaces Treated with Paraloid B72

Xing Zhao [1], Xia Li [1], Siyu Zhang [1], Qing Niu [2], Zongmin Li [3] and Cheng Xue [1,*]

1 School of Cultural Heritage, Northwest University, Xi'an 710100, China; zhaoxing@nwu.edu.cn (X.Z.); lixia@stumail.nwu.edu.cn (X.L.); zhangsiyu1@stumail.nwu.edu.cn (S.Z.)
2 Xi'an Cultural Heritage Promotion Centre, Xi'an 710001, China; niuqing903@163.com
3 Xuzhou Museum, Xuzhou 221000, China; 19952128619@163.com
* Correspondence: w19912992w@163.com

Abstract: In the conservation of cultural relics, the application of Paraloid B72 in humid environments frequently results in the surface whitening of artifacts, which impairs their appearance and conceals important details. This study investigates the mechanisms underlying this phenomenon by examining the effect of ambient humidity, Paraloid B72 mass concentration, solution addition volume, and solvent type. To evaluate the microstructure, transmittance, and composition of the films, a range of analytical techniques were employed, including optical microscopy, scanning electron microscopy, a UV-Visible Spectrophotometer, and Fourier transform infrared spectroscopy. The findings indicate that higher ambient humidity, lower Paraloid B72 mass concentration, smaller solution addition volume, and solvents with higher volatility and water miscibility increase water content during curing, intensifying the whitening effect. These factors modify the interaction between water and solvent, altering the surface structure of Paraloid B72. The whitening mechanism is linked to the cooling effect of solvent volatility, which lowers the dew point temperature at the air–solution interface, causing moisture condensation. Moisture forms droplets that leave irregular pores upon volatility, resulting in surface roughness, optical heterogeneity, and a reduced refractive index, resulting in whitening. This study provides a theoretical basis for understanding and preventing the whitening of Paraloid B72.

Keywords: Paraloid B72; whitening mechanism; condensation

1. Introduction

Protective materials form the material foundation of cultural heritage conservation, playing a crucial role in cleaning, consolidating, bonding, and coating artifacts. Among these materials, acrylic resins are among the most widely used, with Paraloid B72 being particularly prominent [1]. Paraloid B72 is a binary hydrophilic copolymer of ethyl methacrylate (70%) and methyl acrylate (30%) [2,3], appearing as transparent granules soluble in organic solvents such as acetone, ethyl acetate, and toluene. The glass transition temperature is 40 °C, and its density is 8.3 lbs/gal (US). Due to its transparency, durability, and non-yellowing properties [4], Paraloid B72 is extensively used as a consolidant, adhesive, and surface coating agent in the conservation of various materials, including ceramics [5], murals [6], metals [7], and wood [8,9]. Despite its significant role in the preservation of cultural heritage and artifacts, it has been observed that Paraloid B72 can cause whitening on the surfaces of artifacts in humid environments [10,11] (Figure 1).

According to a previous literature search, the majority of current research on the whitening of acrylic materials is focused on emulsion-type acrylic protective materials and acrylic-based latex blend films [12,13], with relatively little attention given to solvent-based acrylic materials. Studies on Paraloid B72 materials mainly address their modification [14,15], but there is a lack of research on the whitening mechanisms of artifacts treated with Paraloid B72. This study investigated the whitening phenomenon of Paraloid B72 films by examining factors involved in the recuring system of Paraloid B72, including ambient humidity, Paraloid B72 mass concentration, solution amount, and solvent type. The effects of these factors on the whitening phenomenon were studied using optical microscopy, scanning electron microscopy, a UV-Visible (UV-VIS) Spectrophotometer, and Fourier transform infrared spectroscopy to analyze the microstructure, transmittance, and composition of the films.

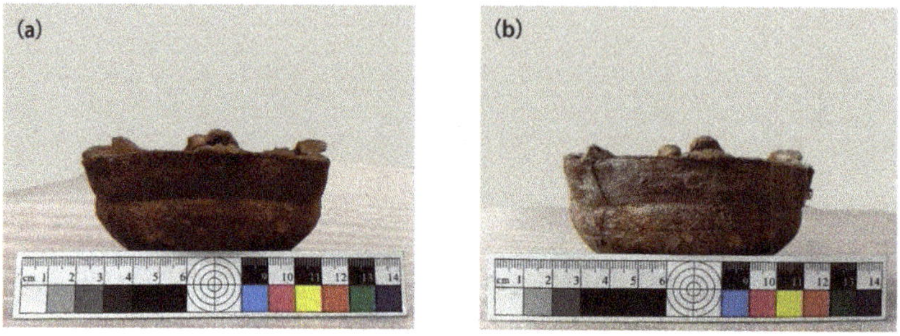

Figure 1. The phenomenon of whitening occurs. (**a**) Before the use of Paraloid B72; (**b**) after the use of Paraloid B72.

The objective of the present study was to contribute to a more comprehensive understanding of solvent-based coating whitening issues by enriching the research on the whitening mechanism of artifact surfaces following the application of Paraloid B72. Furthermore, this research establishes a foundation for future endeavors aimed at elucidating the prevention and restoration of whitening induced by Paraloid B72.

2. Materials and Methods

2.1. Materials and Instruments

Materials: Paraloid B72 (solid, Rohm and Haas, Midland, MI, USA), ethyl acetate (Analytical Reagent, Tianjin Fuyu Fine Chemical Co., Ltd., Tianjin, China), butyl acetate (Analytical Reagent, Macklin Biochemical Co., Ltd., Shanghai, China), acetone (Analytical Reagent, Lianlong Bohua Pharmaceutical Chemistry Co., Ltd., Tianjin, China), sodium dodecyl benzene sulfonate (SDBS, Analytical Reagent, Aladdin Chemical Reagent Co., Ltd., Shanghai, China), and silicone paper.

Instruments: Fourier transform infrared spectrometer (TENSOR 27, Bruker, Billerica, MA, USA), UV-VIS spectrophotometer (Hitachi U2001, Tokyo, Japan), optical microscope (CX40M, Sunny Optical Technology (Group) Co., Ltd., Yuyao, China), and tungsten filament scanning electron microscope (VEGA 3XMU, TESCAN, Brno, Czech Republic).

2.2. Experimental Methods

(1) Sample Preparation: Prior to the formal experiment, our research group conducted a series of preliminary experiments and reached the conclusion that the whitening of Paraloid B72 was related to the environmental humidity during the curing process. Based on this conclusion, we took the participating factors in the curing process as variables to study whether the participating factors would affect the degree of whitening. According to the pre-experiment results, the factors in the curing process of Paraloid B72 were designed as experimental variables to explore the whitening mechanism of Paraloid B72, as shown in Table 1. A certain amount of Paraloid B72 solution was placed in a silicone paper box with a bottom area of 7.6×2.5 cm^2 and cured in a constant temperature and humidity chamber at different ambient humidities, with the temperature set at 25 °C. In order to increase the rationality of the experiment, three parallel samples were set for each group of samples, and the properties of the films were subsequently measured after curing. In order to guarantee the stipulated

humidity conditions and to minimize the impact of the chamber's air circulation system on the curing of Paraloid B72 films, an open acrylic box was positioned within the chamber, with the opening positioned away from the strongest airflow. The samples were then subjected to curing within the aforementioned acrylic box.

Table 1. Specific parameters of the samples.

Sample	Humidity (%)	Paraloid B72 Concentration (%)	Solution Amount (mL)	Solvent	Additive
1	60	10	8	Ethyl acetate	/
2	70	10	8	Ethyl acetate	/
3	80	10	8	Ethyl acetate	/
4	90	10	8	Ethyl acetate	/
5	90	1	8	Ethyl acetate	/
6	90	5	8	Ethyl acetate	/
7	90	15	8	Ethyl acetate	/
8	90	20	8	Ethyl acetate	/
9	90	10	4	Ethyl acetate	/
10	90	10	12	Ethyl acetate	/
11	90	10	8	Ethyl acetate	SDBS
12	90	10	8	Butyl acetate	/
13	90	10	8	Butyl acetate	SDBS
14	60	1	8	Acetone	/

Sodium dodecyl benzene sulfonate (SDBS).

(2) Transmittance Testing: The transmittance of the films was measured using a UV-VIS spectrophotometer. Samples were cut into 0.5 × 0.5 cm pieces and numbered, and then random sampling was conducted for testing the transmittance at 580 nm, with a scanning rate of 100 nm/min. The measurements were taken five times, and the average value was recorded.

(3) Optical Microscopy Testing: The cured Paraloid B72 films were placed on the stage of an optical microscope with the side in contact with silicone paper facing downward. The samples were illuminated with transmitted light, and their morphology was observed at different magnifications.

(4) Infrared Absorption Spectroscopy Testing: The infrared absorption spectra of the films were measured using a Fourier transform infrared spectrometer with an attenuated total reflectance (ATR) attachment. A small amount of film was placed on the ATR sample stage, and the spectra were recorded in the wavenumber range of 4000–600 cm^{-1} with a resolution of 4 cm^{-1}. A total of 64 background and sample scans were conducted.

(5) Scanning Electron Microscopy Testing: The microstructure of the cross-sections of the film samples, fractured in liquid nitrogen, was observed using a scanning electron microscope. The testing conditions were as follows: no gold coating, direct observation, scanning mode: secondary electrons (SEs), accelerating voltage: 10 kV, working distance: 19.95 mm, beam intensity: 10.00.

3. Results and Discussion

3.1. Physicochemical Characteristics of Whitening

To elucidate the underlying causes of the whitening phenomenon in Paraloid B72 films, both the transparent films during the curing process and the whitened films were subjected to microscopic observation and infrared absorption spectroscopy. The microscopic examination of the film's optical micrographs (Figure 2a) revealed a bubble-free surface on the transparent films, with the only imperfections being scratches at the bottom due to contact with silicone release paper. In contrast, the whitened films exhibited irregularly sized pores at the whitened areas, with smaller pores accumulating around larger ones (Figure 2b). The scanning electron microscopy of the cross-sections of the whitened films indicated that pores were present on the side of the film exposed to air and the film surface was rough (Figure 2c). Infrared absorption spectroscopic analysis showed identical absorption spectra for both transparent and whitened films, with no new absorption peaks emerging (Figure 2a), and no new functional group formation showed that the whitening process is a physical change in the phase transition. These findings indicate that the whitening of the films was associated with the formation of pores, and no chemical transformation occurred during the whitening process.

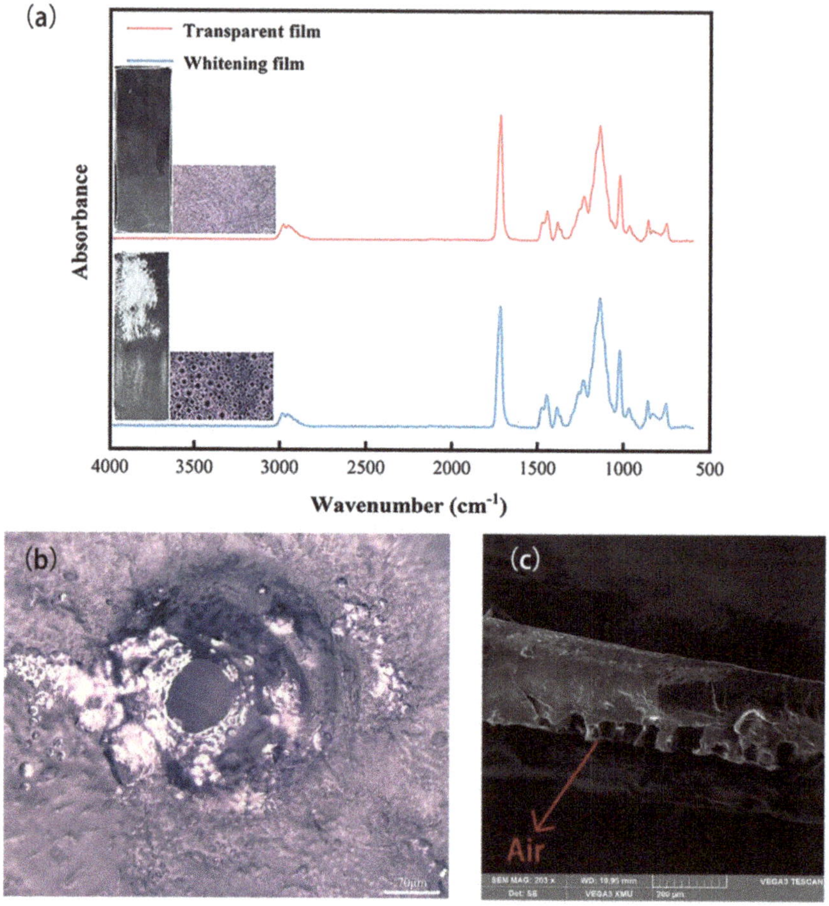

Figure 2. (**a**) Infrared absorption spectrum analysis of transparent film and whitened film; (**b**) optical microscopy at pores; (**c**) pan-white film section microscopy.

3.2. The Effects of Ambient Humidity on Whitening Phenomenon

In practical applications, it has been observed that the Paraloid B72 commonly displays whitening under humid environments. To enhance the differentiation in whitening phenomenon among samples, this study focused on the role of ambient humidity as a variable in conditions of consistent humidity to high humidity. Samples 1–4 were utilized to investigate the effects of varying ambient humidity levels—60%, 70%, 80%, and 90%—on the whitening of Paraloid B72 films. These experiments were conducted with the solvent being ethyl acetate, Paraloid B72 mass concentration at 10%, and a solution volume of 8 mL. Transmittance, a fundamental property of a coating's reflecting appearance, was measured to characterize the degree of whitening and allow for visual comparison. This was accomplished using a UV-VIS Spectrophotometer, which measures film transmittance at the median wavelength of the visible spectrum, specifically at 580 nm [16–18]. The experimental results show that there were variations in the whitening regions among different samples. In this study, we utilized the software ImageJ (1.54j) to calculate the area of the whitening regions. The extent of whitening was quantified by the proportion of the whitening area relative to the total area.

The results indicate that as ambient humidity increased, the area of the whitening region of the film expanded and its transmittance decreased, resulting in a more pronounced degree of whitening. At 60% humidity, the Paraloid B72 film exhibited transparency with a transmittance of 85.30% and a smooth surface devoid of pores. At 70% humidity, the film exhibited a noticeable whitening effect, accompanied by a significant reduction in transmittance to 68.46%. The whitened surface displayed a profusion of dense micro-pores with an average diameter of approximately 8 μm, distributed in a clustered pattern. At 80% humidity, the film exhibited severe whitening, with a further decrease in transmittance to 57.94%. The average pore diameter was approximately 12 μm, and the pores appeared in aggregations resembling islands. At the highest humidity tested (90%), the film exhibited the most pronounced whitening, with a corresponding decrease in sample transmittance to 52.90% (Figure 3). The surface pore density exhibited a notable increase, accompanied by considerable variations in pore size, spanning from approximately 25 μm to 250 μm. The pores exhibited a high degree of independence, with smaller pores coalescing toward larger ones.

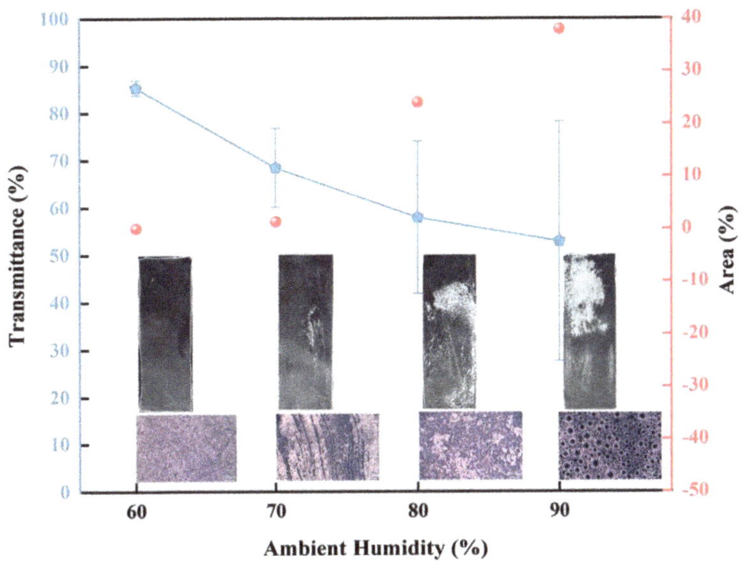

Figure 3. The whitening behavior changes with the ambient humidity (60%, 70%, 80%, 90%) of the environment.

The data indicate that the hypothesis that the occurrence of whitening could be closely related to the moisture content in the air, with the severity of whitening attributed to the increased number of surface pores on the film.

3.3. *Effect of Paraloid B72 Mass Concentration and Solution Amount on Whitening Phenomenon*

The selection of the Paraloid B72 mass concentration in conservation work is contingent upon the underlying material. It is generally used at higher concentrations of 5% and above for consolidation purposes, and at lower concentrations ranging from 1% to 5% for protective coatings [8,19–22]. To reasonably represent the impact of concentration on the properties of Paraloid B72 films, Samples 4 and 5–8 used ethyl acetate as the solvent, with 8 mL of the solution added, to study the effects of Paraloid B72 mass concentrations of 1%, 5%, 10%, 15%, and 20%. In order to enhance the differences in whitening degree among samples, an experimental condition of 90% ambient humidity was selected.

The results indicate that as the mass concentration of the Paraloid B72 increased, the area of the whitening region of the film decreased, while its transmittance increased. At a 1% Paraloid B72 mass concentration, the film's transmittance at 580 nm was 29.44%, with pore diameters between 8 and 1250 μm, composed of densely packed small pores. At a 5% Paraloid B72 mass concentration, the film's transmittance was 18.92%, with pore diameters approximately 50–500 μm, featuring large pores formed by the aggregation of smaller, more uniformly sized and tightly connected pores. However, the whiteness of the films with 5%–20% Paraloid B72 mass concentration decreased with the increase of the concentration. This was likely due to the fact that the transmittance in UV-VIS spectroscopy was also influenced by the thickness of the film, a 1% Paraloid B72 film was too thin, resulting in a higher transmittance compared to a 5% Paraloid B72 film (Figure 4). Thus, an analysis of the relationship between pore diameter and whitening degree revealed that the 1% Paraloid B72 film exhibited greater pore size variability, while the 5% Paraloid B72 film had more uniform pores, indicating a more severe whitening effect from a microscopic perspective.

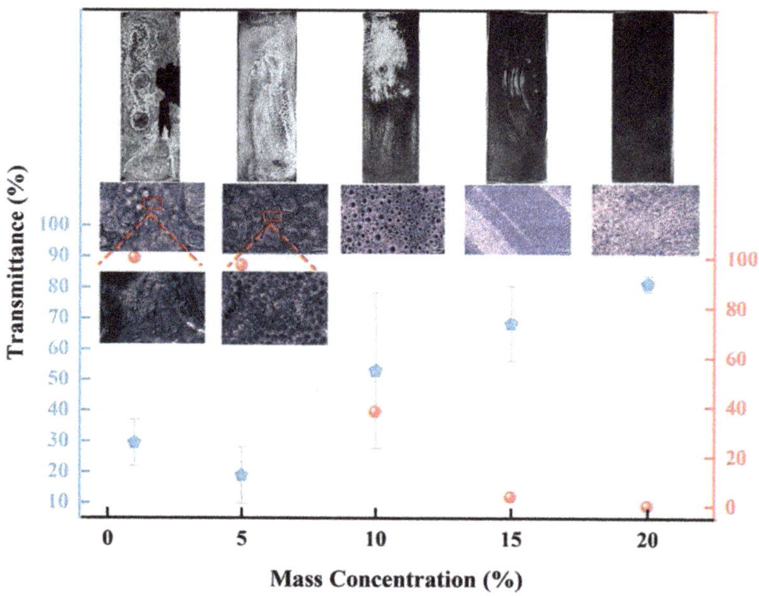

Figure 4. The whitening behavior changes with the Paraloid B72 mass concentration (1%, 5%, 10%, 15%, 20%).

The transmittance of Paraloid B72 films decreased with increasing Paraloid B72 mass concentration, illustrating a correlation between Paraloid B72 concentration and the whitening phenomenon observed in films.

To further explore whether the amount of solution affects the whitening phenomenon, Samples 4, 9, and 10 investigated the influence of 90% ambient humidity, using ethyl acetate as the solvent, with a 10% Paraloid B72 mass concentration and solution volumes of 4 mL, 8 mL, and 12 mL. The experiments demonstrated that, with other variables constant, different solution volumes resulted in distinct whitening degrees after curing. Less solution led to more pronounced whitening, indicating that the degree of whitening was contingent upon the solvent content (Figure 5). Microscopic analysis revealed that the surface pores of film V4 were tightly connected and exhibited a relatively uniform diameter, with an average of approximately 30 μm. In contrast, the surface pores of film V8 exhibited significant variation in diameter, ranging from 25 μm to 250 μm. Additionally, the pores of film V8 were relatively independent, with a tendency for small pores to coalesce into larger ones. In contrast, the pores of film V12 exhibited greater uniformity in size, with an average diameter of approximately 16 μm to 83 μm. Moreover, the pores of film V12 were completely independent. This indicated that the microscopic state of whitening films involved an increase in pore numbers. In this context, the degree of whitening was correlated with either the presence of unevenly sized pores or the presence of uniformly distributed, closely packed pores.

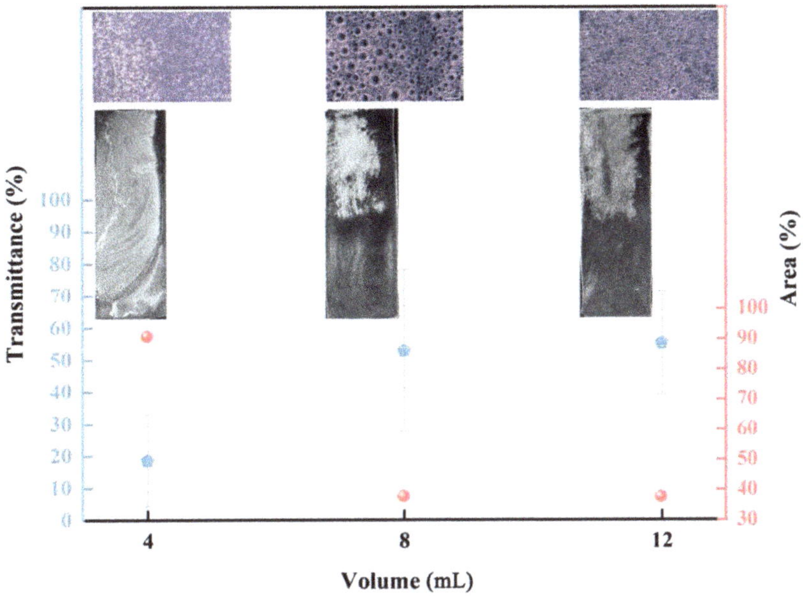

Figure 5. The whitening behavior changes with the addition amount (4 mL, 8 mL, 12 mL).

The principle of solvent evaporation posits that a reduction in solution volume will result in a more rapid overall solvent evaporation rate, thereby facilitating the formation of a film in a shorter time frame. This led to the conclusion that the whitening phenomenon was influenced by the solvent.

3.4. The Effect of Solvent on the Whitening Phenomenon

Since the solvent was the main component of the solution, to investigate its impact on whitening, Samples 4 and 11–13 explored the whitening phenomenon under 90% ambient humidity, with 8 mL of solution added and 10% Paraloid B72 mass concentration, using ethyl acetate and butyl acetate as solvents.

Table 2 shows the physical parameters of the ethyl acetate and butyl acetate. To investigate the effect of solvent miscibility with water and relative evaporation rate on Paraloid B72 film whitening under similar water solubility conditions, 1 g of surfactant was added to the Paraloid B72 ethyl acetate and Paraloid B72 butyl acetate solutions, shaken uniformly, and the upper homogeneous solution was collected for film formation. The experiments showed that, under the same water solubility conditions, the Paraloid B72 ethyl acetate film, with a higher relative evaporation rate, exhibited more severe whitening and had a transmittance of 11.30%, while the Paraloid B72 butyl acetate film with a slower relative evaporation rate had a transmittance of 37.06%. However, contradicting the previous conclusion, the percentage of whitened area of the Paraloid B72 butyl acetate film with higher light transmittance was rather larger. It can be inferred that this phenomenon occurred due to the fact that the low relative volatility of butyl acetate is less affected by machine wind than that of ethyl acetate, resulting in a more uniform whitening effect.

Table 2. Solvent-related physical coefficients.

Solvent	Relative Evaporation Rate	Miscibility with Wate (15 °C)
Butyl acetate	100	7.83%
Ethyl acetate	615	0.50%

In the same solvent, the Paraloid B72 ethyl acetate film exhibited a transmittance of 52.90%, which then sharply declined to 11.30% following the addition of the surfactant. This part was again the same as the previous conclusion—the lower the transmittance of the film, the greater the percentage of whitened area—and the enhanced whitening observed following the addition of the surfactant provided further evidence that the source of whitening involved the participation of atmospheric moisture. From the optical micrographs, it can be seen that the surface of the Paraloid B72 butyl acetate film showed roughness but no holes appeared, and after the addition of SDBS, dense holes appeared, but the diameter of the holes was smaller than that of the Paraloid B72 ethyl acetate +SDBS film. (Figure 6a).

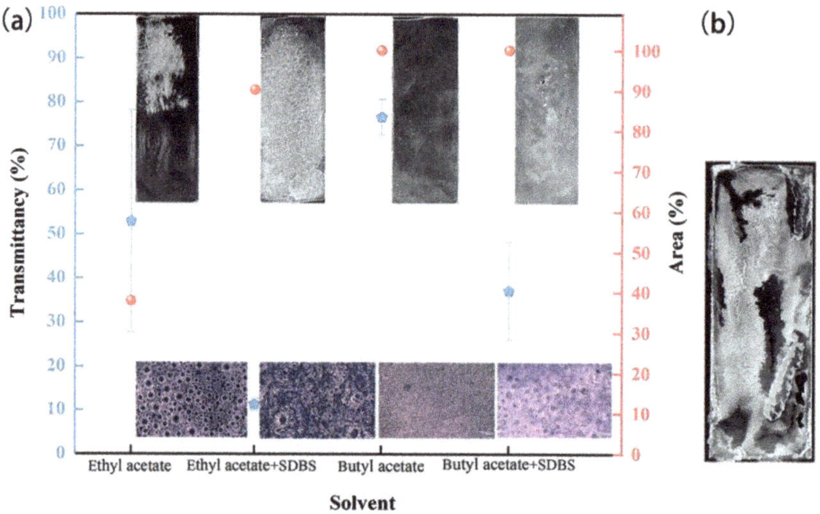

Figure 6. (**a**) Whitening behavior is affected by solvent; (**b**) Paraloid B72–acetone whitening film at 60% ambient humidity.

These experiments demonstrated that a higher solvent water solubility and higher relative evaporation rate led to greater moisture involvement, resulting in the formation of more pores and a more pronounced whitening effect. To verify this result, acetone with a relative evaporation rate of 1120 and infinitely miscible with water was chosen. When a 1% Paraloid B72 acetone solution was cured under 60% humidity, the Paraloid B72 film exhibited whitening even under normal humidity conditions (Figure 6b). This indicated that the whitening phenomenon in Paraloid B72 materials did not have a single critical humidity value, but rather, it was influenced by a complex interplay of factors, including the mass concentration of Paraloid B72, the ambient humidity, the amount of solution, and the solvent.

3.5. Mechanism of Whitening in Coating Films

In this paper, the white degree of the film is characterized by the proportion of the white area and the transmittance. However, it is important to note that film thickness affects the transmittance of the film by the UV-VIS spectrophotometer. This study did not directly investigate the relationship between film thickness and transmittance. Instead, it investigated the factors affecting film thickness, the mass concentration of Paraloid B72, and the volume of solution added separately to explore the influence of each single factor. In accordance with Lamberbier's law, thickness is one of the factors affecting the transmittance, exhibiting a negative correlation. In Samples 1–4, the experimental variables were the different ambient humidity. Theoretically, the thickness of the samples was held constant; however, the experimental results indicate that the higher the humidity, the more pronounced the whitening effect. This suggests that the thickness of the samples did not exert a significant influence on transmittance. In Samples 4, 5–8, the mass concentration of Paraloid B72 was the variable, and the sample thickness increased with the concentration. However, the experimental results indicate that the higher the concentration, the higher the transmittance. The thickness was found to be positively correlated with transmittance, indicating that the thickness had an effect on transmittance, albeit to a lesser degree than that caused by the difference in the solvent relative evaporation rate. In Samples 4, 9, and 10, the variable was the volume of solution added. It was observed that as the volume increased, the film thickness increased, and the transmittance decreased. However, the results also show that as the volume increased, the transmittance increased, indicating that the thickness itself has an influence on the transmittance. However, the degree of influence is considerably less than that caused by the different rates of solvent relative evaporation rate. Similarly, Samples 4 and 12 examined the impact of solvent type on transmittance. The thicker Paraloid B72 butyl acetate film exhibited higher transmittance than the Paraloid B72 ethyl acetate film, suggesting that film thickness may influence transmittance. However, the observed effect was found to be considerably less pronounced than that attributed to differences in the solvent relative evaporation rate (Figure 7).

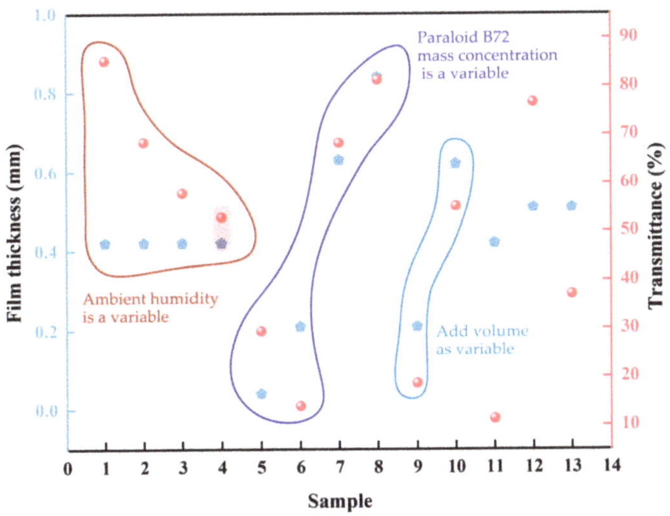

Figure 7. The relationship between the thickness of the sample film and the transmittance.

During the drying and film formation process of coatings, the internal solvent evaporates, causing the particles to closely pack and deposit onto the substrate, resulting in the formation of a solid-state film [23]. The evaporation of the solvent from the surface leads to the shrinkage of the film, creating a porous skeleton. As drying progresses, the gas–liquid interface recedes into the solute particle skeleton, forming menisci within the pores between solute particles. The menisci generate capillary forces, driving the solvent upward to the film surface, thereby emptying most of the solvent from the lower part. The remaining solvent clusters within the skeletal structure evaporate from the inter-particle pores and diffuse to the film surface. As the drying process continues, the particles become increasingly bonded, ultimately forming a pure solid-state dry film [24].

The drying dynamics significantly influence the morphology of the film during the deposition process [25]. The drying kinetics of polymer films can be divided into two stages. In the first stage, the solvent mass diminishes rapidly due to the high solvent content, facilitating rapid solvent transport within the film. The drying behavior is controlled by solvent transport in the gas phase, with the solvent mass fraction at the air/film interface remaining high (>0.25), resulting in significant diffusion and evaporation. As the solvent evaporates and its load in the film reduces, solvent transport slows, and drying becomes increasingly governed by solvent diffusion within the film and gas–liquid phase equilibrium. In the second stage, although the solvent concentration at the air/film interface remains lower (<0.25), diffusion slows, preventing adequate regeneration of the solvent near the interface. Consequently, the solvent content at the gas–liquid interface approaches zero, significantly reducing the drying rate. The diffusion coefficient is dependent on the polymer/solvent ratio, and as the solvent evaporation flux weakens, the concentration gradient is reduced [26–31].

In the gaseous environment of the film drying and formation system, in the absence of alterations in water vapor content and air pressure, a reduction in temperature results in the air reach saturation with water vapor. If the temperature continues to decline below the dew point, the water vapor will condense into liquid water droplets [32]. During the condensation process, water vapor molecules randomly aggregate on the condensation surface, forming micron-sized clusters that grow until initial nucleation occurs. The initial liquid droplets absorb surrounding water vapor in order to grow. Initially, adjacent small droplets appear isolated, but they subsequently couple in temperature fields, with significant fluid field interference between droplets. This leads to the formation of observable liquid bridges. These bridges extend under the influence of surface tension to form larger, more stable droplets during the late stage of coalescence (Figure 8) [33].

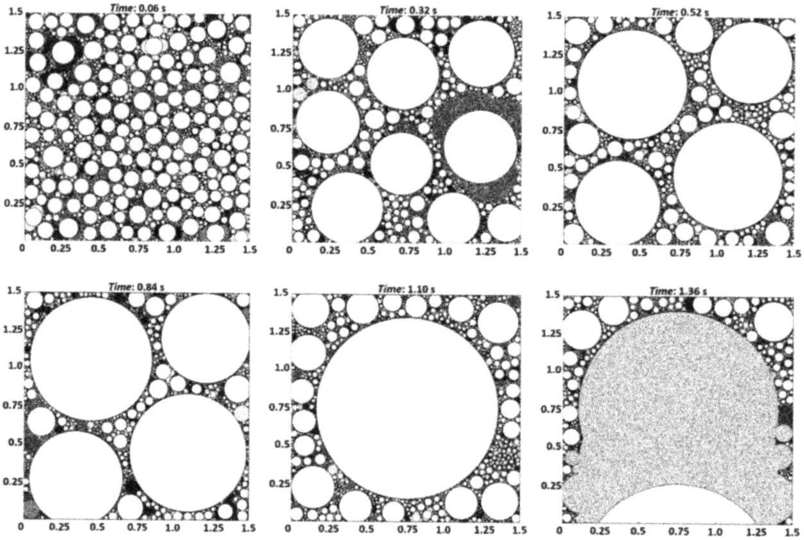

Figure 8. Visualization of the nucleation and growth of droplets during dropwise condensation of pure steam [33].

In materials science, it is observed that polymer composites experience stress whitening under tensile, compressive, and bending stresses. Researchers have found that whitening in different materials is due to cavities formed under stress, leading to changes in the refractive index, presenting as white [34–36]. This study demonstrates that the appearance of holes causes a reduction in the refractive index, resulting in whitening.

The experimental results indicate that during the curing process of Paraloid B72 films, rapid solvent evaporation in the first stage absorbs heat, lowering the air/film interface temperature. As the temperature continues to decline below the dew point, the water in the surrounding air becomes saturated and begins to condense on the film surface. As the surface viscosity of Paraloid B72 increases and solvent evaporation slows into the second stage, the Paraloid B72 films cure, water starts to evaporate, and the cavities left by the evaporating water droplets appear as observed holes (Figure 9). Infrared absorption spectroscopy analysis of Paraloid B72 film samples indicates that the observed whitening is caused by physical changes due to alterations in surface roughness, which manifest as holes.

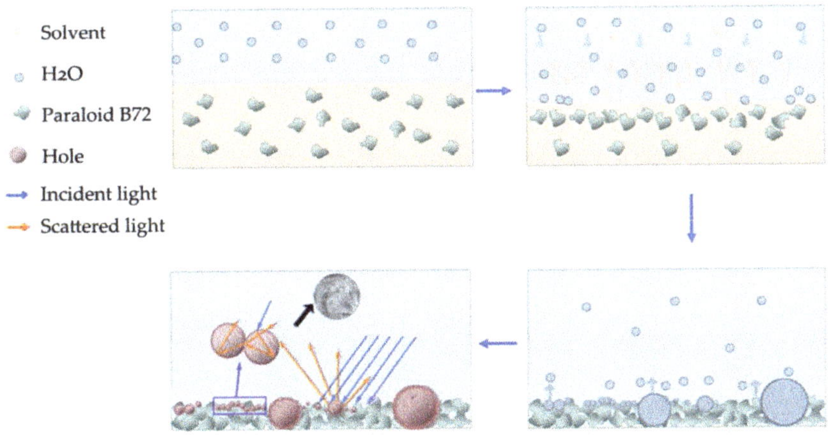

Figure 9. Paraloid B72 film whitening mechanism diagram.

The extent of film surface whitening during curing is influenced by a number of factors, including ambient humidity, the concentration of Paraloid B72, the quantity of solution added, and the type of solvent used. These factors exert a direct or indirect influence on the interaction between water and the film surface, which in turn leads to an increased involvement of water in condensation. Sufficient water leads to closer distances between droplets, causing smaller droplets to merge into larger ones. A lower Paraloid B72 mass concentration results in higher solvent content, prolonging the first stage of solvent evaporation, making nearby air more prone to condensation. As the Paraloid B72 mass concentration increases, particle packing on the Paraloid B72 surface facilitates reaching the second stage of solvent evaporation, extending the overall drying time, causing droplets to invade the hydrophilic Paraloid B72 structure. However, higher concentrations shorten the first stage of solvent evaporation, restricting the solvent evaporation rate and reducing condensation. Lower solution addition amounts significantly shorten the drying time with the same gas–liquid interface area, increasing water content in the air, causing droplets to contact. With increased solution addition, small droplets aggregate, and new small droplets form due to surface tension. Further increases in solution addition result in greater Paraloid B72 surface deposition, affecting the solvent evaporation rate and reducing the first stage drying time, thereby impacting condensation. The type of solvent directly reflects the solvent's miscibility with water and its relative evaporation rate. A higher miscibility and evaporation rate facilitate whitening; under the same experimental conditions, the surface of the Paraloid B72 butyl acetate film showed a rough state, but no holes appeared

(Figure 10). The solvent's evaporation absorbs heat, causing water in the air to condense on the curing film surface, participating in film curing and forming irregular cavities on the surface. These cavities lead to optical inhomogeneity and a reduction in the refractive index, resulting in whitening.

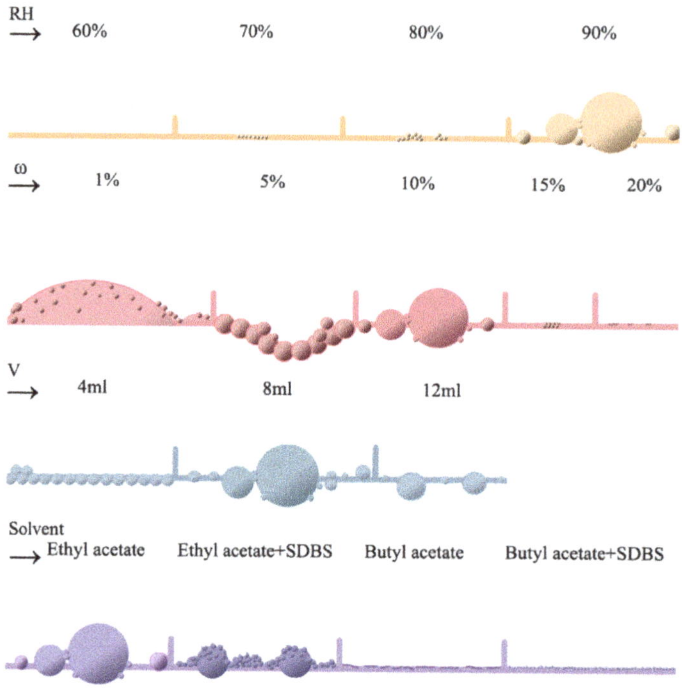

Figure 10. Microscopic scheme of holes on the film surface.

4. Conclusions

This study investigated the whitening mechanism of Paraloid B72 films by examining the effects of ambient humidity, Paraloid B72 mass concentration, solution addition volume, and solvent type during the curing process. The key findings are as follows:

(1) Moisture content is a crucial factor in determining the extent of whitening. Variations in influencing factors lead to increased moisture content in the air, a decrease in the dew point temperature at the gas–liquid interface, moisture condensation on the film surface, and pore formation upon evaporation. This surface roughness causes optical heterogeneity, resulting in whitening.

(2) During the Paraloid B72 curing process, factors such as ambient humidity, Paraloid B72 mass concentration, solution addition volume, and solvent type are interrelated. There is no fixed threshold for ambient humidity. Under normal humidity conditions, managing other influencing factors can still produce a whitening effect on the film.

(3) In routine cultural heritage conservation, controlling factors such as ambient humidity (it is recommended to use under 50% ambient humidity) and solvent type can minimize the moisture involved in condensation or prevent the microenvironment from reaching the dew point temperature, thereby preventing whitening. According to the conclusion that the whitening is due to the change in physical structure on the surface of Paraloid B72, a further study could investigate whether the Paraloid B72 film can revert to the transparent state after the physical structure is changed again by

placing the white film in an environment with a temperature higher than the glass transition temperature.

Author Contributions: X.Z.: investigation and writing—original draft. X.L.: investigation and writing—review. S.Z.: investigation and writing—review. Q.N.: investigation. Z.L.: writing—review and editing. C.X.: revising, funding, and supervision. All authors have read and agreed to the published version of the manuscript.

Funding: This research was funded by the National Natural Science Foundation of China (52203126), the National Social Science Foundation of China (22FKGB009), the Key Research and Development Plan of Shaanxi Province (2022SF321), the Projects of the Social Science Foundation of Shaanxi Province (2021G001), the Shaanxi Province Youth Science and Technology Rising Star Project (2024ZC-KJXX-033), and the National Cultural Relics Protection Special Fund Project (22-5-13-3200-314).

Institutional Review Board Statement: Not applicable.

Informed Consent Statement: Not applicable.

Data Availability Statement: The datasets used and/or analyzed during the current study are available in a publicly accessible.

Conflicts of Interest: The authors declare no conflicts of interest.

References

1. Centenaro, S.; Franceschin, G.; Cattaruzza, E.; Traviglia, A. Consolidation and Coating Treatments for Glass in the Cultural Heritage Field: A Review. *J. Cult. Herit.* **2023**, *64*, 132–143. [CrossRef]
2. Chiantore, O.; Lazzari, M. Characterization of Acrylic Resins. *Int. J. Polym. Anal. Charact.* **1996**, *2*, 395–408. [CrossRef]
3. Chiantore, O.; Lazzari, M. Photo-Oxidative Stability of Paraloid Acrylic Protective Polymers. *Polymer* **2001**, *42*, 17–27. [CrossRef]
4. Podany, J.; Garland, K.M.; Freeman, W.R.; Rogers, J. Paraloid B-72 as a Structural Adhesive and as a Barrier within Structural Adhesive Bonds: Evaluations of Strength and Reversibility. *J. Am. Inst. Conserv.* **2001**, *40*, 15–33. [CrossRef]
5. Elghareb, W.K. Analysis and Restoration of Greco-Roman Pottery in Egypt. *Int. J. Conserv. Sci.* **2023**, *14*, 1327–1342. [CrossRef]
6. Zhou, X.; He, W.; Ou, J.; Hu, Y.; Wang, F.; Fang, X.; Li, W.; Amirfazli, A. Translucent Superhydrophobic Coating for Murals Protection. *Colloids Surf. A Physicochem. Eng. Asp.* **2024**, *689*, 133750. [CrossRef]
7. Han, Z.; Huang, X.; Chen, J.; Chen, J. 2-Mercaptobenzimidazole Compounded with the Conventional Sealer B72 for the Protection of Rusted Bronze. *J. Cult. Herit.* **2024**, *67*, 42–52. [CrossRef]
8. Soytürk, E.E.; Kartal, S.N.; Terzi, E.; Önses, M.S.; Şarkdemir, K.; Çelik, N. Evaluation of Wood Treated with Paraloid B72® and Boric Acid: Thermal Behavior, Water Absorption and Mold Resistance. *Eur. J. Wood Wood Prod.* **2023**, *81*, 923–934. [CrossRef]
9. Ionescu, C.Ş.; Lunguleasa, A.; Avram, A.; Spîrchez, C. Evaluation of the Efficiency of the Consolidation Treatment with Paraloid B72, Performed on Artworks with Degraded Wood Support. *EDP Sci.* **2021**, *343*, 02001. [CrossRef]
10. Han, K.; Teri, G.; Cheng, C.; Tian, Y.; Huang, D.; Ge, M.; Fu, P.; Luo, Y.; Li, Y. Evaluation of Commonly Used Reinforcement Materials for Color Paintings on Ancient Wooden Architecture in China. *Herit. Sci.* **2024**, *12*, 122. [CrossRef]
11. Molina, M.T.; Cano, E.; Leal, J.; Fort, R.; Álvarez de Buergo, M.; Ramírez-Barat, B. Protective Coatings for Metals in Scientific—Technical Heritage: The Collection of the Spanish National Museum of Science and Technology (MUNCYT). *Heritage* **2023**, *6*, 2473–2488. [CrossRef]
12. Stafford, O.; Hinderliter, B.; Croll, S. Electrochemical Impedance Spectroscopy Response of Water Uptake in Organic Coatings by Finite Element Methods. *Electrochim. Acta* **2006**, *52*, 1339–1348. [CrossRef]
13. Gong, X.; Han, X.; Chen, K. Anti-Aging Performance Evaluation of Acrylate Emulsion Used for Cultural Relics Conservation. *Spectrosc. Spectr. Anal.* **2023**, *43*, 2181–2187. [CrossRef]
14. Mohamed, E.H. Effectiveness Assessment of Paraloid B-72 Enhanced with Nano Materials to Improve of Completion Mortars Properties for Conservation of Seti I Temple in El-Qurna, Thebes West Bank, Egypt. *Multiscale Multidiscip. Model. Exp. Des.* **2023**, *6*, 371–387. [CrossRef]
15. He, W.; Ou, J.; Wang, F.; Lei, S.; Fang, X.; Li, W.; Amirfazli, A. Transparent and Superhydrophobic Coating via One-Step Spraying for Cultural Relic Protection against Water and Moisture. *Colloids Surf. A Physicochem. Eng. Asp.* **2023**, *662*, 130949. [CrossRef]
16. Zhao, J.; Wang, Y.; Liu, C. Film Transparency and Opacity Measurements. *Food Anal. Methods* **2022**, *15*, 2840–2846. [CrossRef]
17. Yousefi, F.; Mousavi, S.B.; Heris, S.Z.; Naghash-Hamed, S. UV-Shielding Properties of a Cost-Effective Hybrid PMMA-Based Thin Film Coatings Using TiO2 and ZnO Nanoparticles: A Comprehensive Evaluation. *Sci. Rep.* **2023**, *13*, 7116. [CrossRef]
18. Yao, Y.; Zhang, Z.; Wang, K.; Fu, Q. Effects of Plasticizing on Mechanical and Viscous Characteristics of Poly (Vinyl Alcohol): A Comparative Study between Glycerol and Diethanolamine. *Macromol. Mater. Eng.* **2023**, *308*, 2300090. [CrossRef]
19. Abdelbar, M.A. Technical Investigation and Conservation of Thin-walled Core Cast Leaded Bronze Statuettes of Osiris. *Shedet* **2024**, *12*, 325–343. [CrossRef]

20. Avram, A.; Ionescu, C.S.; Lunguleasa, A. A Consolidation of Degraded Lime Wooden Support from Heritage Objects Using Two Types of Consolidant. *BioResources* **2023**, *18*, 4580. [CrossRef]
21. El-Sayed, S.S.M. Restoration and Conservation of an Archaeological Marble Gravestone from the Greek Era-Al-Fustat-Egypt. *J. Herit. Des.* **2024**, *4*, 338–351.
22. Abu-Baker, A. Analytical Investigation and Electrochemical Conservation Treatment for Archaeological Copper Alloy Artifacts from Jordan. *Conserv. Património* **2023**, *42*, 38–55. [CrossRef]
23. Zhang, B.; Fan, B.; Huang, Z.; Higa, K.; Battaglia, V.S.; Prasher, R.S. A Review of Dispersion Film Drying Research. *J. Electrochem. Energy Convers. Storage* **2023**, *20*, 030801. [CrossRef]
24. Holl, Y.; Keddie, J.; McDonald, P.; Winnik, W. Drying Modes of Polymer Colloids. *Film. Form. Coat.* **2001**, *790*, 2–26. [CrossRef]
25. Weng, K.; Ye, L.; Zhu, L.; Xu, J.; Zhou, J.; Feng, X.; Lu, G.; Tan, S.; Liu, F.; Sun, Y. Optimized Active Layer Morphology toward Efficient and Polymer Batch Insensitive Organic Solar Cells. *Nat. Commun.* **2020**, *11*, 2855. [CrossRef]
26. Guerrier, B.; Bouchard, C.; Allain, C.; Bénard, C. Drying Kinetics of Polymer Films. *AIChE J.* **1998**, *44*, 791–798. [CrossRef]
27. Merklein, L.; Eser, J.C.; Börnhorst, T.; Könnecke, N.; Scharfer, P.; Schabel, W. Different Dominating Mass Transport Mechanisms for Drying and Sorption of Toluene-PMMA Films–Visualized with Raman Spectroscopy. *Polymer* **2021**, *222*, 123640. [CrossRef]
28. Börnhorst, T.; Scharfer, P.; Schabel, W. Drying Kinetics from Micrometer-to Nanometer-Scale Polymer Films: A Study on Solvent Diffusion, Polymer Relaxation, and Substrate Interaction Effects. *Langmuir* **2021**, *37*, 6022–6031. [CrossRef]
29. Ronsin, O.J.; Jang, D.; Egelhaaf, H.-J.; Brabec, C.J.; Harting, J. A Phase-Field Model for the Evaporation of Thin Film Mixtures. *Phys. Chem. Chem. Phys.* **2020**, *22*, 6638–6652. [CrossRef]
30. Vinçotte, A.; Beauvoit, E.; Boyard, N.; Guilminot, E. Effect of Solvent on PARALOID® B72 and B44 Acrylic Resins Used as Adhesives in Conservation. *Herit. Sci.* **2019**, *7*, 42. [CrossRef]
31. Yu, J.; Shen, Z.; Lu, W.; Zhu, Y.; Liu, Y.-X.; Neher, D.; Koch, N.; Lu, G. Composition Waves in Solution-Processed Organic Films and Its Propagations from Kinetically Frozen Surface Mesophases. *Adv. Funct. Mater.* **2023**, *33*, 2302089. [CrossRef]
32. He, X.; Wang, S.; Zhang, B. A Semi-Theoretical Model for Water Condensation: Dew Used in Conservation of Earthen Heritage Sites. *Water* **2020**, *13*, 52. [CrossRef]
33. Mirafiori, M.; Tancon, M.; Bortolin, S.; Del Col, D. Modeling of Growth and Dynamics of Droplets during Dropwise Condensation of Steam. *Int. J. Heat Mass Transf.* **2024**, *222*, 125109. [CrossRef]
34. Farahani, M.F.; Bagheri, R.; Marouf, B.T. Investigation on the Onset and Progress of Stress Whitening in Polypropylene Using Digital Image Processing. *Polym. Bull.* **2024**, *81*, 7139–7156. [CrossRef]
35. Zhang, S.; Bhagia, S.; Li, M.; Meng, X.; Ragauskas, A.J. Wood-Reinforced Composites by Stereolithography with the Stress Whitening Behavior. *Mater. Des.* **2021**, *206*, 109773. [CrossRef]
36. Lyu, D.; Tang, Y.; Qian, L.; Chen, R.; Lu, Y.; Men, Y. Large Strain Cavitation Induced Stress Whitening in Propylene-Butene-1 Copolymer during Stretching. *Polymer* **2019**, *167*, 146–153. [CrossRef]

coatings

MDPI

Article

Assessing the Fire Properties of Various Surface Treatments on Timber Components in Ancient Chinese Buildings: A Case Study from the Xianqing Temple in Changzhi, Shanxi, China

Yupeng Li [1], Sokyee Yeo [1,*], Weihan Zou [1] and Shibing Dai [2,3,*]

1 School of Human Settlements and Civil Engineering, Xi'an Jiaotong University, Xi'an 710049, China; lyp3120122010@stu.xjtu.edu.cn (Y.L.); zouweihan78@stu.xjtu.edu.cn (W.Z.)
2 Key Laboratory of Ecology and Energy Saving Study of Dense Habitat, Ministry of Education, Shanghai 200092, China
3 College of Architecture and Urban Planning, Tongji University, Shanghai 200092, China
* Correspondence: yeosokyee@mail.xjtu.edu.cn (S.Y.); daishibing@tongji.edu.cn (S.D.)

Abstract: Traditional and modern coatings play a key role in enhancing the fire resistance of ancient Chinese buildings. However, further comparative analysis is needed on the fire properties of the two coatings and their effects on different timber structural components. This study focuses on the main hall of the Shanxi Changzhi Xianqing Temple, a typical traditional column and beam construction built between the Song and Jin periods. Firstly, the combustion characteristics of various timber structural component samples with different surface treatments (traditional "Yi-ma-wu-hui" and modern flame retardants) were analyzed using cone calorimeter. Secondly, the fire development process of the Xianqing Temple building model was analyzed by a fire dynamics simulator (FDS), and the effect mechanism of different surface treatments on the burning process was further studied. The results show that the fire resistance of timber structural components is significantly improved after modern and traditional surface treatments. The traditional method is more effective in delaying the peak heat release rate and reducing the surface temperature during combustion, while the modern surface treatment significantly prolongs the ignition time of the timber structural components. The FDS results confirm that modern and traditional surface treatments significantly improve the fire resistance of the building, delaying the flashover time by about 300 s, with no collapse occurring within 800 s. In addition, the fire resistance of buildings after traditional surface treatment is better compared to traditional methods. The above research results can provide direct data support for the selection and optimization of fireproof coatings and treatment methods for ancient buildings.

Keywords: ancient Chinese buildings; timber structural components; fire properties; surface treatments; cone calorimeter; FDS

1. Introduction

In ancient Chinese architecture, improving the fire resistance of buildings has always been an important consideration. Traditionally, applying mud plaster and using brick walls have been some of the fire prevention measures that have been commonly adopted. In particular, Yi-ma-wu-hui, a fireproof traditional coating that was formed by mixing raw hemp with lime plaster and other organic additives (such as tung oil, pig blood, and paper pulp), was often applied to the surface of timber structures to enhance the fire properties of the building. This traditional surface treatment was emphasized in the "Principles for the Conservation of Wooden Built Heritage" published by the 19th International Council on Monuments and Sites (ICOMOS) in 2017 [1]. In recent years, the application of modern flame-retardant coatings has gradually increased, providing more possibilities for fire prevention in ancient timber structural buildings [2,3]. New flame retardants include various inorganic or organic chemical agents that can significantly improve the flame retardancy of wood. Compared to traditional methods, these chemical flame retardants have the advantages of simple preparation and long-lasting effects, making them very suitable for the fire prevention treatment of timber structures in large-scale or difficult-to-disassemble ancient buildings [4,5]. For example, in key parts of timber structures such as mortise and tenon connections, timber column beam frames, the application of transparent flame retardants through full-surface brushing can strengthen the fire resistance capability of these timber structures, reduce the risk of fire, and have a minimal visual impact [6,7].

Previous studies have revealed the main components and content ratios of the "Yi-ma-wu-hui" used in Chinese timber structures of ancient buildings [8,9] and confirmed that this technique can significantly enhance the fire properties of timber components [10–12]. "Yi-ma-wu-hui" can increase the fire resistance limitation of timber components to a certain extent and reduce fire destructions. Also, the traditional material provides a protective layer at high temperatures, insulating oxygen and reducing heat transfer [8]. In addition, the favorable application effects of various modern flame retardants in the fire prevention properties of timber structures in ancient buildings have also been confirmed in previous studies. These modern flame retardants often contain intumescent components that form a foam layer at high temperatures, insulating and stopping the flame from spreading. Moreover, a good-quality fireproofing coating can significantly extend the ignition time and fire resistance of wood [13–15]. Compared to the original timber structure components, the heat release rate of components treated with flame retardants has been reduced by 53.1% [16,17]. The addition of flame retardants also reduces the concentration of smoke and carbon dioxide gas released, lowering the fire scene's temperature [6]. However, although the above studies have confirmed that both the traditional "Yi-ma-wu-hui" prevention technique and flame retardants can achieve better fire prevention properties in ancient buildings, further comparative analyses on the fire prevention effects of these surface treatments and their effects on different timber structural components are still needed.

Past studies on the fire resistance of timber structural components in ancient Chinese buildings revealed that the cone calorimeter has been widely used for evaluating the fire properties of these components [18]. This analytical technique could simulate a real-time fire environment by exposing the timber structural components to stable heat radiation, allowing for the precise measurement and analysis of the material's reaction to heat. Through testing and analysis with the cone calorimeter, important data on how wood and fire prevention treatment materials react under different heat radiation intensities can be obtained, which subsequently help to optimize fire treatment techniques and explore the fire resistance properties of various types and ratios of flame retardants [19,20]. Additionally, due to the high cost of actual fire tests and the limited data collected, the fire dynamics simulator (FDS) has gradually become an important tool in the study of fire properties in timber structures of ancient buildings [21,22]. FDS uses parallel computing to simulate the temporal and spatial distribution of parameters including smoke, temperature, and oxygen concentration, accurately depicting the fire development process. Researchers can design various complex geometric scenarios and fire source models according to the real environment, simulating fire scenarios under different ventilation conditions, burning materials, and layouts. In addition, previous studies have revealed that the fire resistance properties of timber structures in ancient buildings are closely related to parameters using FDS, such as the fire source position, the timber structure types, and the window area of buildings [23–31].

Based on the above analysis, this study focuses on the traditional column and beam construction of the Shanxi Changzhi Xianqing Temple main hall that was built between the Song and Jin dynasties, with the aim of exploring the fire performance of different woods and surface treatments. Elm, pine, and poplar wood, which are widely used for the roof rafter of the main hall of the Xianqing Temple, were taken as the objects of this study. Firstly, traditional "Yi-ma-wu-hui" and modern flame retardants are used for the surface treatment of different types of wood samples (elm, pine, and poplar). Next, the combustion characteristics of the wood samples under different surface treatment conditions are analyzed using the cone calorimeter, and the fire resistance properties of wood samples with different surface treatments are studied comparatively. Finally, based on the combustion characteristic data of different wood samples, a PyroSim model of the Xianqing Temple main hall is established using FDS for fire dynamic simulation analysis, revealing the effect of different surface treatments on the fire development process of the building.

2. Materials and Methods

2.1. Materials

Built between the Song and Jin dynasties, Xianqing Temple is located in Shangcun village, Changzhi City, Shanxi Province of China. The main timber structural components of the temple main hall are primarily constructed of poplar wood (*Populus* L.) (Figure 1). According to previous research, the upper part of the building and the areas near the fire source usually experienced the highest temperatures, and smoke that did not dissipate in time tended to accumulate around the roof rafter region, thus making the roof rafters the most vulnerable area within the timber frame [26]. Therefore, in this study, the rafter component of the Xianqing Temple was chosen as the study object. On-site investigations revealed that some timber structural components, particularly the rafters, were replaced with pine (*Pinus sylvestris* var. *mongolica* Litv.) and elm (*Ulmus pumila* L.) from previous restorations (Figure 2); hence, three wood species, namely elm, pine, and poplar wood, were selected for testing.

Figure 1. Exterior and interior images of the Xianqing Temple.

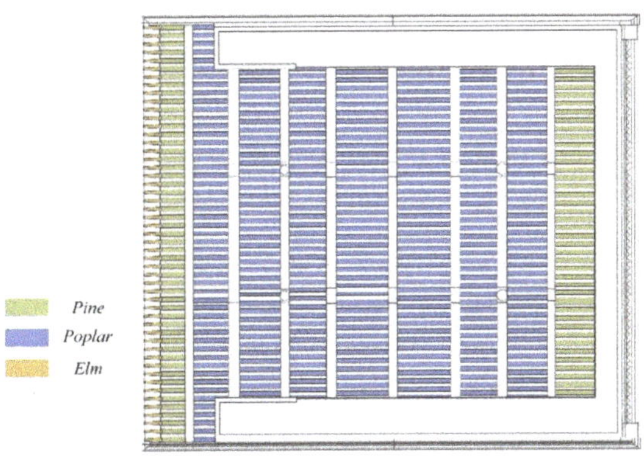

Figure 2. Timber species statistics of roof rafter in the main hall of Xianqing Temple.

Based on the above findings, a total of 9 samples, consisting of 3 samples for each of the three wood species—elm, pine, and poplar—were prepared. Due to the size limitation of the cone calorimeter tray, the test samples were specifically designed to be 90 mm × 90 mm × 6 mm so that they could fit into the tray accurately. All samples were maintained at a moisture content of less than 18% and were then subjected to surface polishing prior to the application of surface treatments. Two surface treatments were applied to the three types of wood samples. Treatment 1 involved the application of modern flame retardant (Remmers Adolit BSS 1, boron-free water-based), where the flame retardant was applied thinly with a brush, air-dried, and repeated for the second time to ensure that two coatings were applied. Treatment 2 involved the application of the traditional "Yi-ma-wu-hui" technique (detailed processing can be found in our previous findings [32]), where the main ingredients included lime plaster, hemp, tung oil, pig blood, and paper pulp. In addition, a control group with no surface treatment was set up for comparison purposes. Detailed sample information can be found in Tables 1 and 2, and the sample processing flowchart is shown in Figure 3.

Table 1. Detailed sample information.

Timber Species	Density (g/cm^3)	Surface Treatment	Specimen Number
Elm	0.680	No treatment (control)	E-N
		Modern	E-R
		Traditional	E-T
Pine	0.423	No treatment (control)	Pi-N
		Modern	Pi-R
		Traditional	Pi-T
Poplar	0.386	No treatment (control)	Po-N
		Modern	Po-R
		Traditional	Po-T

Note: E—elm; Pi—pine; Po—poplar; N—no treatment; T—traditional treatment; R—retardant.

Table 2. Average moisture content of timber samples.

Timber Species	Average Moisture Content before Treatment (%)	Specimen Number	Average Moisture Content after Treatment (%)	Moisture Content Variation (%)
Elm	10.2	E-N	16.9	+6.7
		E-R	8.7	−1.5
		E-T	8.7	−1.5
Pine	7.6	Pi-N	14.4	+6.8
		Pi-R	6.9	−0.7
		Pi-T	7.4	−0.2
Poplar	8.6	Po-N	10.7	+2.1
		Po-R	8	−0.6
		Po-T	8.2	−0.4

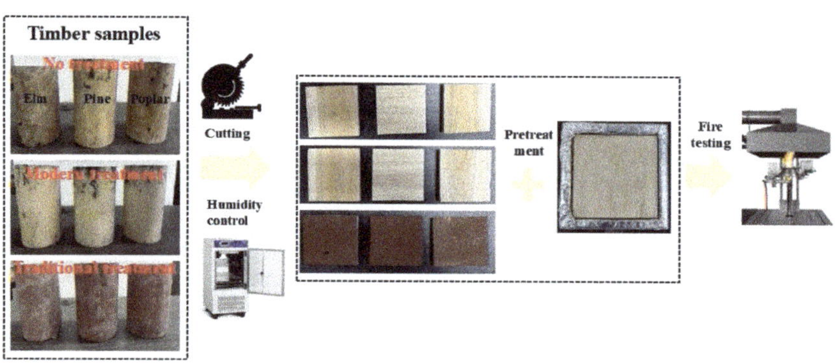

Figure 3. Timber sample processing flowchart.

2.2. Methods

2.2.1. Cone Calorimeter

The model of the cone calorimeter experimental instrument is FTT0242 (Fire Testing Technology, East Grinstead, UK). Before conducting the combustion test on the samples, it is essential to ensure that there are no impurities inside the sample combustion chamber and to keep it clean and dry. This step is crucial because any residue left inside the chamber may cause the samples to melt and detach unevenly during the combustion process, which might later affect the authenticity of the data and lead to potential deviations in the experimental results. During the test, the samples were placed in the sample chamber and tightly wrapped with aluminum foil to prevent heat loss and external interference, ensuring that this went directly to the heater. Additionally, to ensure test accuracy, it is necessary to maintain a distance of 25 mm between the sample and the cone heater. In this study, a heat flux value of 25 kW/m^2 was determined experimentally and used for testing. To prevent deformation such as curling of the wood samples during heating and combustion, an iron grid was added above the wood samples to secure the wood and ensure the stability of the experiment. During the experiments, standards (methanol combustion) were used to calibrate the thermofluid meter and ensure that the environmental conditions (temperature and relative humidity) were consistent for each experiment.

2.2.2. Fire Dynamics Simulation

In this experiment, fire dynamics simulation (FDS) software is used for modeling the ancient building and further simulating its fire process. Due to the software's incompatibility with curved objects, cylindrical components such as columns and rafters in the building could not be built directly. Hence, the complex components in this model are stacked by rectangular objects. Based on the research object rafter size of 100 mm in this study, the model mesh size is set to 100 mm × 100 mm × 100 mm, which is shown to run normally in the software page. Considering the actual situation of the Xianqing Temple, the fire source was set near the altar during the modeling process. In the simulation process, the wind speed was set to 2 m/s (northwest direction), the ambient temperature was 293.15 K, the atmospheric pressure was 103 kPa, and the specific heat capacity of air was 1.0069 kJ/(kg·K). The combustible material type of the ancient building was stacked wood, so the value of α was set to 0.0469 kW/s^2, the maximum heat release rate Q was 8.0 MW, and the simulation time was 800 s. For combustion property parameters that could not be measured in the experiment (such as conductivity), all specimens used existing data for pines from the material library, with a thermal conductivity of 0.1 W/(m·K), absorption coefficient of 5 × 104 m^{-1}, radiation coefficient of 0.9, and ignition temperature of 300 °C (Table 3). The combustion properties of wood under different surface treatment conditions were determined through cone calorimeter experiments.

Table 3. Combustion property parameters of fire dynamics simulation.

Properties	Data
Density	0.1 W/(m·K)
Absorption coefficient	5 × 10^4 m^{-1}
Emissivity	0.9
Ignition	300 °C

After setting the basic parameters required for the fire dynamics simulation, the FDS software (Version 6.8.0) can then be used to analyze the fire development in the building. In the simulation, three temperature monitoring points and three temperature monitoring profiles were set (Figure 4). The monitoring points were located in the air to ensure the validity of the data, with specific positions as follows: next to the fire source at coordinates (point 1: 6.5, 11.5, 1.8) and at the center of the ridge at coordinates (point 2: 6.0, 7.0, 8.3); on the roof directly above the fire source at coordinates (point 3: 6.0, 11.5, 5.9). The coordinates of the temperature monitoring profiles are Y = 4.0 m, Y = 7.0 m, and Y = 10.0 m.

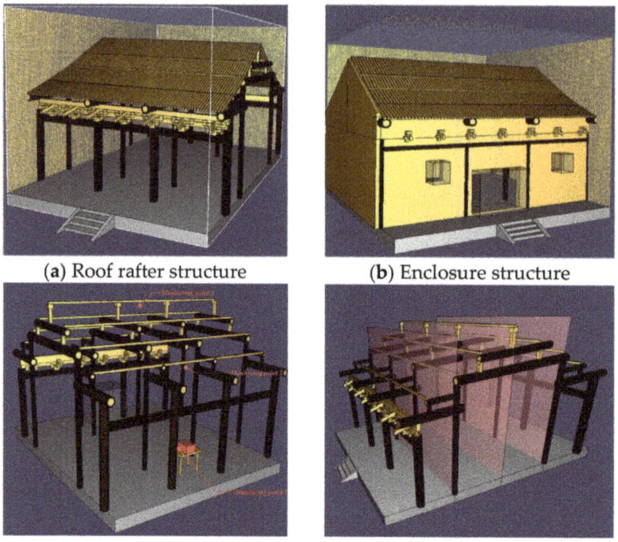

(**a**) Roof rafter structure (**b**) Enclosure structure

(**c**) Monitoring point location (**d**) Profile Location

Figure 4. Schematic of FDS building model, temperature monitoring points, and profile locations.

3. Results

3.1. Cone Calorimeter Analysis

Time to ignition (TTI) measures the time required for a material to transition from the start of heating to sustain combustion under a certain intensity of incident heat flux. A longer TTI implies better fire resistance of the material. Experimental data (Table 4) show that the TTI for elm, pine, and poplar wood is 80 s, 61 s, and 52 s, respectively. Therefore, among these three types of wood samples, poplar is the most flammable, while elm wood exhibits the best fire resistance. In addition, wood samples treated with modern surface treatments demonstrate significant improvements in fire resistance, with TTI values for elm, pine, and polar increasing to 145 s, 128 s, and 103 s, representing percentage increases of 81.3%, 109.8%, and 98.2%, respectively. The results have been confirmed by previous studies on the fire performance of surface treatments on pine and beech wood [33]. Compared to modern surface treatments, traditional treatments (Yi-ma-wu-hui) are equally effective, resulting in TTI values for elm, pine, and polar increasing to 131 s, 127 s, and 102 s, with percentage increases of 63.8%, 108.2%, and 96.2%, respectively. Although slightly less effective than modern treatments, both surface treatments significantly enhance the fire resistance of timber structural components.

Heat release rate (HRR) is a measure of the amount of heat released per unit area during the combustion process of a material under a certain heat flux, used to assess fire intensity. The peak heat release rate (pkHRR) reflects the maximum heat release of the material during combustion and is an important indicator of a fire hazard. As shown in Figure 5, the HRR curves of pine and elm wood exhibit two peaks, with the second peak higher than the first peak. In contrast, the HRR curve of poplar wood initially shows a more gradual increase without a distinct first peak. This difference may be related to the moisture content of the different types of wood. During the initial stages of combustion, moisture in the wood evaporates and produces hydrogen gas (H_2) as a combustible material that participates in the combustion process, leading to differences in the initial peaks of heat release.

Table 4. Time to ignition of wood samples with different surface treatments.

Sample Number	Ignition Time	Combustion Time
E-N	80 s	175 s
E-R	145 s	345 s
E-T	131 s	345 s
Pi-N	61 s	185 s
Pi-R	128 s	300 s
Pi-T	127 s	415 s
Po-N	52 s	180 s
Po-R	103 s	395 s
Po-T	102 s	310 s

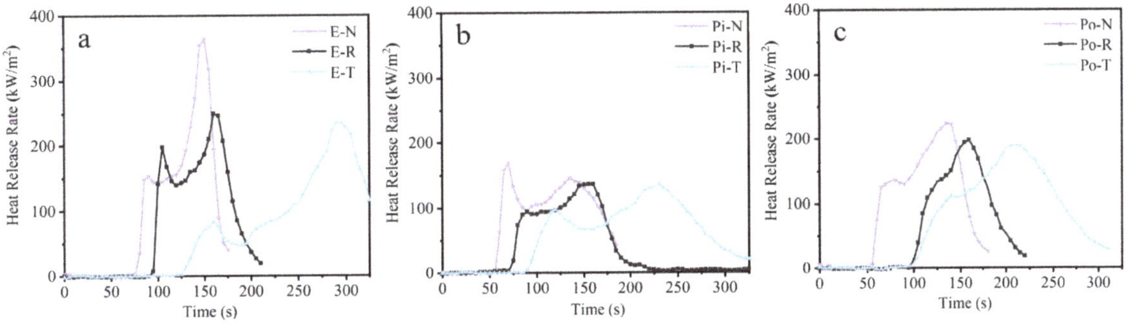

Figure 5. Variation in heat release rate of wood samples: (**a**) elm samples; (**b**) pine samples; (**c**) poplar samples.

Different surface treatment methods have a significant impact on the pkHRR of wood, effectively reducing the peak heat release rate and decreasing the potential danger of flame spread. Modern surface treatment methods have led to a decrease in the pkHRR of elm, pine, poplar, and wood by 31.4%, 19.9%, and 26.4%, respectively (Table 5). The effects of traditional surface treatment methods for elm, pine, and poplar are even more significant, with reductions of 35.3%, 18.1%, and 29.0%, respectively. Additionally, both surface treatment methods suppress heat release in the initial stages of combustion, delaying the occurrence of peaks and prolonging the duration of wood components in a fire, with the traditional method showing a more pronounced delay effect. This indicates that both traditional and modern surface treatment techniques can effectively enhance the safety properties of wood in fire scenarios, especially in terms of delaying flame development and reducing peak heat release, which helps increase response time at the fire scene.

Table 5. Heat release rate and total heat release of wood samples.

Specimen Number	Heat Release Rate Peak (kW/m^2)	Time to Peak (s)	Total Heat Release (MJ/m^2)	Average Heat Release Rate (kW/m^2)
E-N	363.8	150	17.3	182.2
E-R	250.0	160	16.2	76.7
E-T	235.5	290	23	110.2
Pi-N	169.7	70	14.4	115.1
Pi-R	136.0	155	11.8	39.4
Pi-T	136.0	230	18.8	58.5
Po-N	224.9	135	17.8	157.8
Po-R	197.0	160	13.5	116.1
Po-T	189.4	210	23.1	112.1

Mass loss rate (MLR) refers to the rate of change in sample mass over time during the combustion process, reflecting the extent of volatilization and combustion of the material under a certain fire intensity. The MLR of the three types of wood samples is shown in Figure 6, where it can be observed that the mass of the specimens changes slowly within the first 100 s of combustion, sharply decreases between 100 and 250 s, and then stabilizes after 250 s. In contrast, the mass loss of the samples applied with surface treatments is not as pronounced after 200 s, and the change in mass loss is similar for both modern and traditional surface treatment methods. This result was confirmed in previous investigations of the fire performance of different types of cedar after surface treatment [34]. To provide a clearer representation of the rate of mass loss and understand the variation in mass loss during the combustion process, this study also recorded the mass loss rate curves for the wood samples at different time points and plotted their fitting curves (Figure 7). Due to the nonlinear variation in the mass loss rate with time curve, a polynomial curve fitting was performed to combine the actual variation in the curve and the fitting effect. In addition, since the mass change curves of the three types of wood are similar, the analysis will focus on the fitting curve of the mass loss rate over time for elm wood. The results generally show an initial trend of decrease, followed by an increasing trend, and returning back to a decreasing trend. This indicates that wood samples undergo rapid mass loss in the early stages of combustion, with the mass loss rate gradually increasing, reaching a peak before decreasing again. In addition, it is evident that the peak time for the rate of mass loss in wood samples is significantly delayed after using modern and traditional surface treatments, with the effect being more pronounced in the traditional surface treatments.

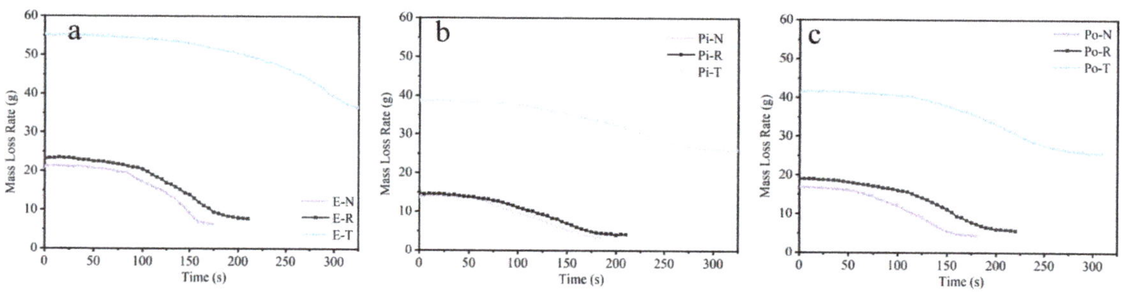

Figure 6. Mass change curves with time of wood samples: (**a**) elm mass change curve; (**b**) pine mass change curve; (**c**) poplar mass change curve.

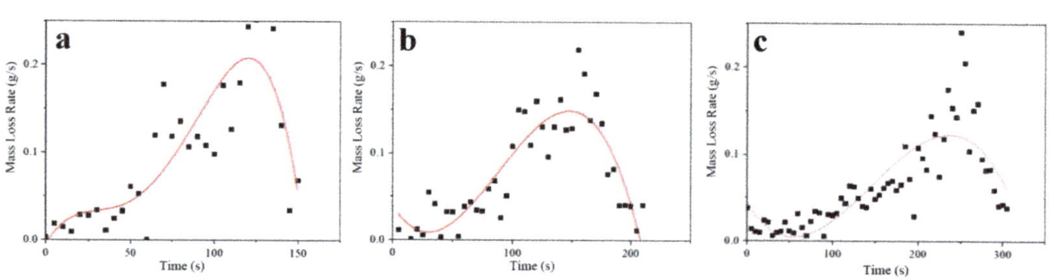

Figure 7. Curves of sample mass rate versus time: (**a**) untreated (R^2 = 0.73897); (**b**) modern treatment (R^2 = 0.77118); (**c**) traditional treatment (R^2 = 0.67378).

Surface treatment alters the rate of the charring of wood, resulting in lower mass loss for untreated wood compared to wood that has undergone surface treatment within the same burning time. Although there is no consistent pattern in the total mass loss of wood samples after complete combustion, wood that has been surface-treated tends to have a longer burning time, with the peak mass loss rate occurring significantly later than in untreated wood samples (Table 6). The results confirm that surface treatments can effectively delay the rate of wood combustion, thereby increasing its durability in a fire, and that traditional surface treatments are more effective.

Table 6. Mass loss rate of wood samples with different surface treatments.

Specimen Number	Loss of Specimen Mass under Conical Calorimeter (g)	Loss of Specimen Mass under One-Sided Fire Combustion (g)	MLR (g/s)	Time to Peak Mass Loss Rate (s)
E-N	13.4	7.6	0.27	141
E-R	15.5	5.7	0.21	151
E-T	18.1	---	0.24	238
Pi-N	9.9	19.7	0.15	93
Pi-R	17.6	5.9	0.13	153
Pi-T	12.7	5.7	0.15	117
Po-N	12.5	25.3	0.16	96
Po-R	10.4	10.1	0.20	113
Po-T	15.0	11.6	0.12	172

Smoke production rate (SPR) is an indicator of the speed at which smoke is generated in a fire, typically expressed as the ratio of the mass of smoke produced per unit of time to the mass of the burning material. The curves show the variation in the smoke production rate for wood samples with different surface treatments (Figure 8). It can be observed that after undergoing modern and traditional surface treatments, the time span for smoke generation significantly increases for elm, pine, and poplar wood, with wood samples treated using traditional methods showing a longer duration of smoke production.

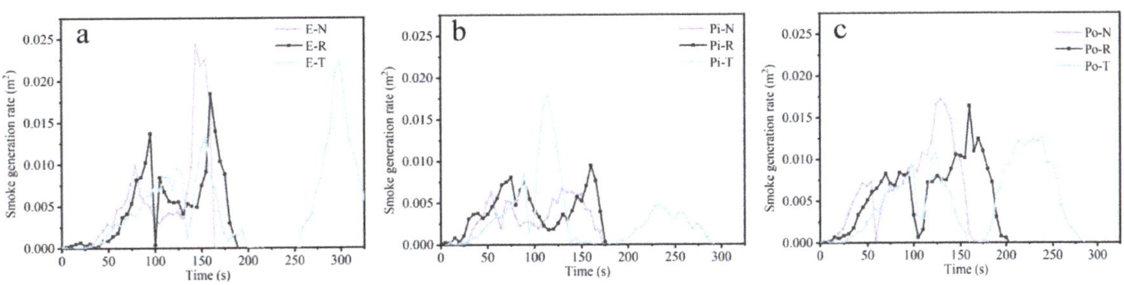

Figure 8. Smoke generation rate curves for wood samples with different surface treatments: (**a**) elm samples; (**b**) pine samples; (**c**) poplar samples.

According to Table 7, it could be found that the total oxygen consumption, total smoke production, total smoke release, total CO release, and total CO_2 release are correlated during wood combustion. The production of CO_2 was higher when the wood was sufficiently combusted, while the production of CO and smoke was lower. CO_2 was positively correlated with oxygen consumption, while CO was negatively correlated with oxygen consumption. In addition, the smoke release is also negatively correlated with CO_2, implying that incomplete combustion results in a relative decrease in CO_2 production. Furthermore, it is evident that the total smoke released during the combustion process significantly increases for various types of wood after modern and traditional surface treatments, with the traditional surface treatment showing the most significant increase. The smoke release totals for elm, pine, and poplar wood increased by 49.3%, 62.2%, and 14.4%, respectively. Furthermore, the completeness of combustion can be assessed by calculating the ratio of carbon monoxide (CO) to carbon dioxide (CO_2) emissions. It can be observed that compared to poplar and elm wood, pine wood releases more CO after application of modern or traditional surface treatments, indicating incomplete combustion and higher fire resistance, with the effect being more pronounced after traditional surface treatment.

Table 7. Smoke release information for wood samples with different surface treatments.

Sample Number	Total Oxygen Consumption (TOD) (g)	Total Smoke Production (m^2)	Total Smoke Release (m^2/m^2)	Total CO Release (kg/kg)	Total CO_2 Release (kg/kg)	CO-Release-to-CO_2-Release Ratio
E-N	11.6	0.75	85.0	0.50	19.1	0.026
E-R	10.8	0.92	104.9	0.74	25.7	0.028
E-T	15.5	1.12	127.0	1.02	50.1	0.021
Pi-N	9.5	0.45	50.7	0.41	29.7	0.014
Pi-R	7.8	0.71	80.7	5.39	40.1	0.134
Pi-T	12.4	0.73	82.8	5.57	12.4	0.449
Po-N	9.0	0.76	86.1	0.52	28.0	0.019
Po-R	11.3	0.97	110.4	0.75	28.0	0.026
Po-T	15.3	1.11	125.1	0.93	49.4	0.018

Overall, among the three types of wood samples without surface treatment, the fire resistance properties from highest to lowest are elm wood, pine wood, and poplar wood. After surface treatment, the fire resistance properties of these three types of wood remain consistent. In the experiment, poplar wood is more prone to combustion compared to the other two types of wood and releases more smoke during the combustion process. Although poplar wood's average heat release rate falls between pine and elm wood, it releases more total heat due to its shorter burning time.

Both modern and traditional surface treatment methods can effectively improve the fire resistance properties of wood samples, as shown in Table 8, where a positive value ("+") indicates a positive effect on the fire resistance properties and a negative value ("−") indicates a negative effect. The combustion property parameters are reflected in the prolonged ignition time, reduced heat release rate, and slowed mass loss rate, which help mitigate the loss of the mechanical properties of wood and the effects of temperature fields during a fire. However, especially with the traditional treatment method, although it delays the peak time of the heat release rate, it also leads to the generation of a large amount of smoke and incomplete combustion of the wood. In contrast, the modern treatment method produces less smoke and demonstrates a more significant advantage in increasing ignition time.

Table 8. Enhancement effect of different surface treatments on fire parameters of wood samples.

Fire Resistance Parameters	Timber Species	Modern	Traditional
Ignition time	Elm	+81.3%	+63.8%
	Pine	+109.8%	+108.2%
	Poplar	+98.2%	+96.2%
Peak heat release rate	Elm	+31.4%	+35.3%
	Pine	+19.9%	+18.1%
	Poplar	+26.4%	+29%
Delay time to peak mass loss rate	Elm	+7.1%	+68.8%
	Pine	+64.5%	+25.8%
	Poplar	+7.3%	+79.2%
Smoke release rate	Elm	−22.4%	−49.2%
	Pine	−59.2%	−63.3%
	Poplar	−14.6%	−47.1%

3.2. Fire Dynamics Simulation Analysis

To further explore the fire resistance properties of buildings under different surface treatment methods, the fire development process of the Xianqing Temple building was simulated using FDS, and the changes in temperature at three monitoring points were recorded, as shown in Figure 9. During the simulation, the temperature at these monitoring points gradually increased, with monitoring point 3 (at 8.3 m) reaching the highest temperature, followed by monitoring point 2 (at 5.9 m), and monitoring point 1 showing the lowest temperature (at 1.8 m). This temperature distribution reflects the behavioral characteristics of smoke and flames in the early stages of a fire in ancient buildings, where temperatures at higher positions inside the building rise earlier and faster compared to lower positions. This is mainly because flames spread along the surface of the combustible material, and due to the upward trend of hot air and thermal currents, flames are more likely to expand upwards. Additionally, the flames heat the surrounding air, creating thermal currents. As the thermal currents rise, cold air is replenished from below, creating a cyclic convection process. Natural convection further promotes the upward diffusion of flames and thermal currents.

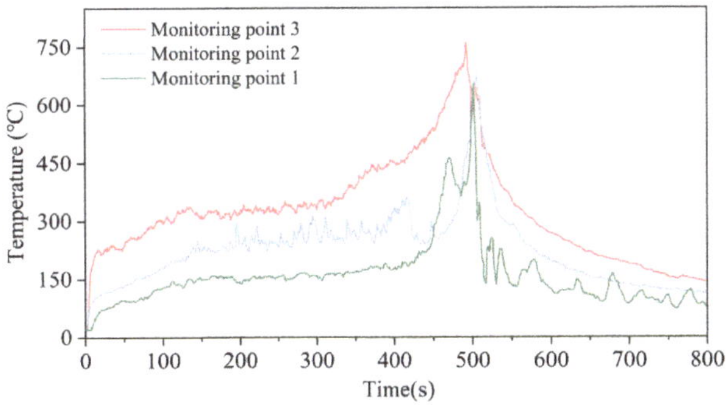

Figure 9. Temperature changes in the main hall of Xianqing Temple (timber structural components without various surface treatments).

Furthermore, the highest temperatures at the three monitoring points reached 760 °C, 653 °C, and 680 °C, respectively. The temperature on the roof rafter quickly reaches 200 °C and stabilizes at around 330 °C, which is the typical charring temperature of wood. When the fire temperature reaches 335 s, the internal temperature of the building rapidly increases, indicating a flashover phenomenon occurring at that time.

Flashover is the most unfavorable scenario in a fire, indicating a sudden full-scale combustion inside the building. In firefighting, it is generally recognized that the period before flashover occurs in a building is the prime time to extinguish the fire to preserve the overall structure of the building. To clearly determine the timing of flashover in a building, the trend of temperature differentials at each moment compared to the previous moment was further plotted to observe the rate of temperature change over time (Figure 10). The moment when the temperature change rate rapidly accelerates can be identified as the time when the building experiences flashover in a fire.

As shown in Figure 10a, the flashover time of the Xianqing Temple roof rafter is approximately 370 s. After 500 s, the temperature differentials at each monitoring point change from positive to negative, indicating the highest temperatures in the building and rapid cooling in the later stages. Based on this phenomenon, it can be inferred that the building collapses at this time, with heat dissipating outward, and the temperatures at each measurement point gradually decrease to 100 °C. Combining the above analysis results, it is confirmed that when a fire breaks out at the Xianqing Temple due to the careless use of fire for sacrificial purposes, the fire should be quickly extinguished within 300 s to preserve the overall building structure.

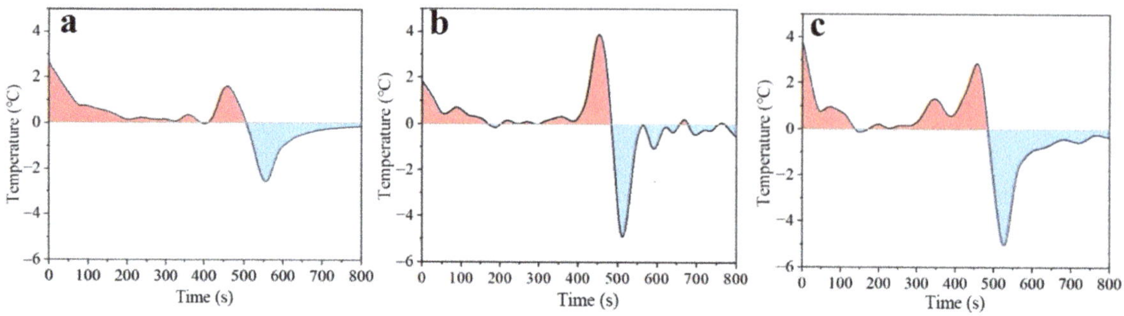

Figure 10. Temperature change rate curves for the main hall of Xianqing Temple (timber structural components without various surface treatments): (**a**) monitoring point 1; (**b**) monitoring point 2; (**c**) monitoring point 3.

To visually illustrate the temperature distribution of the Xianqing Temple roof rafter more intuitively, temperature distribution maps at three profiles over time were plotted (Figures 11–13). In order to facilitate the demonstration of the analyses and findings, during the initial stages of the fire when the temperature was slowly increasing, recordings were taken at intervals of 100 s. However, due to the rapid temperature changes in the later stages when the building experienced flashover, recordings were taken every 20 s.

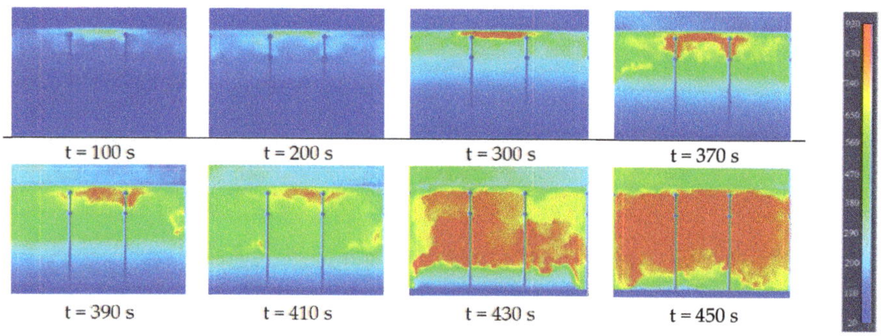

Figure 11. Temperature distribution maps at Y = 10 m profiles of the main hall of Xianqing Temple (timber structural components without various surface treatments).

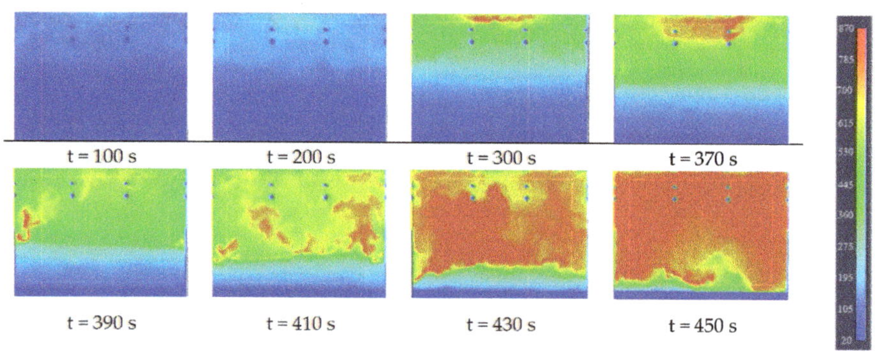

Figure 12. Temperature distribution maps at Y = 7 m profiles of the main hall of Xianqing Temple (timber structural components without various surface treatments).

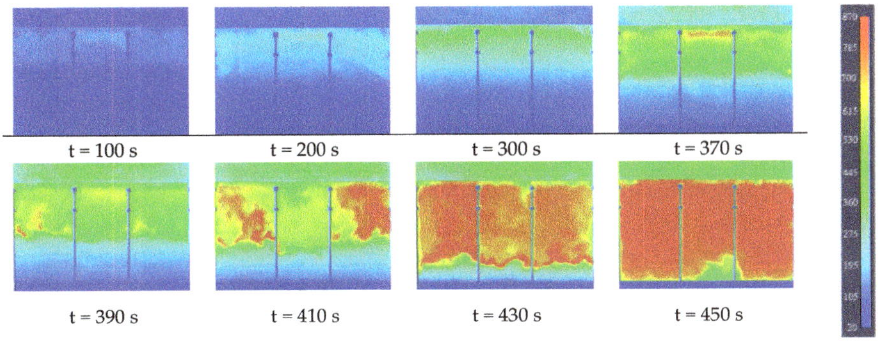

Figure 13. Temperature distribution maps at Y = 4 m profiles of the main hall of Xianqing Temple (timber structural components without various surface treatments).

From Figures 11–13, it can be observed that smoke initially accumulates at the top of the roof rafter inside the building and then gradually spreads to the sides and downward, eventually enveloping the entire building in a high-temperature field. This result further confirms that during a fire at the Xianqing Temple, the roof collapsed first due to the flashover.

After understanding the combustion and temperature changes in the Xianqing Temple, based on the ignition time, heat release rate, mass loss rate, and smoke production of pine, poplar, and elm wood with different surface treatments, the combustion properties of poplar wood with the worst fire resistance were selected as simulation conditions to further study the fire properties of buildings with different surface treatments. The temperature changes at different monitoring points in buildings with different surface treatments are shown in Figure 14.

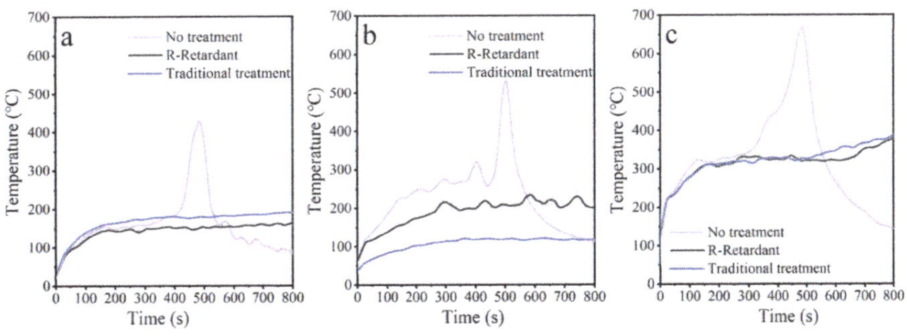

Figure 14. Temperature changes in the main hall of Xianqing Temple with different surface treatments: (**a**) monitoring point 1; (**b**) monitoring point 2; (**c**) monitoring point 3.

It can be observed that the Xianqing Temple main hall model without surface treatment collapsed at 492 s. However, modern and traditional surface treatment methods kept the temperature around 330 °C, with the modern treatment maintaining the collapsed temperature until 500 s after the fire started, while the traditional treatment stabilized at 330 °C until 640 s (monitoring points 3 and 2). At the same time, the temperature change at monitoring point 1, located at a lower position, remained below 200 °C. This result confirms that both modern and traditional surface treatment methods effectively prolong the time before the building ignites. Furthermore, no flashover or collapse occurred within the simulated 800 s, indicating that both modern and traditional treatment methods extended the optimal rescue time for the building by approximately 300 s.

Table 9 records the time taken for monitoring point 3 to reach the wood combustion charring temperature (300 °C). Under the no-treatment scenario, 300 °C charring temperature is reached at 91 s after the fire started, while the traditional and modern surface treatment delayed reaching the charring temperature by 37 s and 43 s, respectively. In summary, both modern and traditional surface treatment methods not only provide good fire resistance properties but also help to delay the destruction of the building and provide a longer rescue time for evacuation and firefighting in case of a fire.

Table 9. Time to reach 300 °C of monitoring point 3 in the main hall of Xianqing Temple with different surface treatments.

Surface Treatment	Time for the Roof (Monitoring Point 3) to Reach 300 °C
No treatment (N)	91 s
Modern (R)	134 s
Traditional (T)	128 s

4. Conclusions

This study analyzed the fire properties of different surface-treated woods such as elm, poplar, and pine using a cone calorimeter. Taking the Xianqing Temple as a case study, the fire development process was simulated using FDS to explore the impact of different surface treatments on the fire resistance properties of traditional timber buildings. The cone calorimeter test results showed that compared to pine and elm wood, poplar wood had the poorest fire properties with the shortest ignition time. When subjected to modern and traditional surface treatments, significant changes were observed in fire property parameters such as peak heat release rate, total heat release, and mass loss rate, significantly improving the fire properties of the timber structural components. The traditional surface treatments were more effective in delaying the peak heat release rate and reducing the surface temperature during combustion, while the modern treatments significantly extended the ignition time of the wood.

The FDS simulation results indicated that after the traditional timber frame building was ignited, smoke first accumulated at the roof rafter of the building, resulting in the highest temperature at the top. The smoke then gradually spread to the sides and downward, eventually enveloping the entire building in a high-temperature field. Furthermore, both modern and traditional surface treatments delayed the flashover time of the Xianqing Temple main hall by approximately 300 s, and no signs of building collapse occurred within 800 s, further demonstrating the significant improvement in the fire properties of the building with these treatments. The results of this study can be directly applied to the restoration and protection of ancient buildings, optimizing fire prevention measures. It also provides guidance for the development of new fire-resistant materials and technologies. Considering the limitations of this study, further investigations will be conducted on other types of wood and surface coatings and different timber structural combinations to analyze their fire performance. The FDS will also be used to simulate the fire transmission paths in complex building structures and thus evaluate the overall fire resistance.

Author Contributions: Conceptualization, Y.L., S.D. and S.Y.; methodology, S.D. and S.Y.; validation and formal analysis, Y.L. and S.Y.; investigation, Y.L. and S.Y.; writing—original draft preparation, Y.L.; writing—review and editing, S.Y. and W.Z.; supervision, S.Y.; funding acquisition, S.D. and S.Y. All authors have read and agreed to the published version of the manuscript.

Funding: This research was funded by the Key Laboratory of Ecology and Energy Saving Study of Dense Habitat, Ministry of Education (Grant No. 20220105) and National Natural Science Foundation of China (NSFC) (Grant No. 52378033).

Data Availability Statement: The original contributions presented in the study are included in the article; further inquiries can be directed to the corresponding author.

Conflicts of Interest: The authors declare no conflicts of interest.

References

1. ICOMOS. *Principles for the Preservation of Historic Timber Structures*; ICOMOS: Paris, France, 2007.
2. Chen, L.; Wang, Y.Z. A review on flame retardant technology in China. Part I: Development of flame retardants. *Polym. Adv. Technol.* **2010**, *21*, 1–26. [CrossRef]
3. Zhang, Y.; You, F.; Zhang, Q. Protection of Historical Buildings by Using NEW chemical Techniques: Taking Nanjing as an Example. In Proceedings of the 7th International Conference on Intelligent Computation Technology and Automation, Changsha, China, 25–26 October 2014; IEEE: Piscataway, NJ, USA, 2015; pp. 436–440.
4. Wang, K.; Wang, S.; Meng, D.; Chen, D.; Mu, C.; Li, H.; Sun, J.; Gu, X.; Zhang, S. A facile preparation of environmentally-benign and flame-retardant coating on wood by comprising polysilicate and boric acid. *Cellulose* **2021**, *28*, 11551–11566. [CrossRef]
5. Mohamed, A.L.; Hassabo, A.G. Flame Retardant of Cellulosic Materials and Their Composites. In *Flame Retardants: Polymer Blends Composites and Nanocomposites*; Springer: Berlin/Heidelberg, Germany, 2015; pp. 247–314.
6. Wang, X.; Wang, J.; Wang, J.; Sheng, G. Experimental and numerical simulation analyses of flame spread behaviour over wood treated with flame retardant in ancient buildings of Fuling Mausoleum, China. *Fire Technol.* **2022**, *58*, 1–25. [CrossRef]
7. Pan, C.; Jiang, X.; Yang, Y.; Fu, Z.; Meng, C.; Xi, S. Research on fireproofing treatments and Properties of ancient timber buildings. *Sci. Conserv. Archaeol.* **2018**, *30*, 28–36.
8. Peifan, Q.; Deqi, Y.; Qi, M.; Aijun, S.; Jingqi, S.; Zengjun, Z.; Jianwei, H. Study and restoration of the Yi Ma Wu Hui layer of the ancient coating on the Putuo Zongcheng Temple. *Int. J. Archit. Herit.* **2021**, *15*, 1707–1721. [CrossRef]
9. Fu, P.; Teri, G.L.; Li, J.; Li, J.X.; Li, Y.H.; Yang, H. Investigation of ancient architectural painting from the Taidong tomb in the western qing tombs, Hebei, China. *Coatings* **2020**, *10*, 688. [CrossRef]
10. Chen, L.Z.; Xu, Q.F.; Han, C.Q.; Wang, Z.C.; Leng, Y.B.; Chen, X. Experimental study on fire endurance of timber beams with treatment of one hemp fiber and five plastering exposed to three-side fire. *J. Build. Struct.* **2021**, *42*, 101–109.
11. Wang, Z.; Xu, Q.; Han, C.; Chen, L.; Chen, X. Experimental study on fire endurance of round timber column with traditional craftwork with one hemp fiber and five plastering application. *J. Build. Struct.* **2017**, *47*, 14–19.
12. Xu, Q.; Han, C.; Chen, L.; Wang, Z.; Leng, Y. Experimental study on mechanical behavior of timber beams treated with traditional plastering application after exposed to three-side fire. *J. Build. Struct.* **2021**, *51*, 92–97.
13. Morgan, A.B. The future of flame retardant polymers–unmet needs and likely new approaches. *Polym. Rev.* **2019**, *59*, 25–54. [CrossRef]
14. Lazar, S.T.; Kolibaba, T.J.; Grunlan, J.C. Flame-retardant surface treatments. *Nat. Rev. Mater.* **2020**, *5*, 259–275. [CrossRef]
15. Osório, M.; Fonseca, E.M.M.; Pereira, D. Fire Resistance in Screwed and Hollow Core Wooden Elements Filled with Insulating Material. *Fire* **2024**, *7*, 288. [CrossRef]
16. Chen, G.; Chen, C.; Pei, Y.; He, S.; Liu, Y.; Jiang, B.; Jiao, M.; Gan, W.; Liu, D.; Yang, B.; et al. A strong, flame-retardant, and thermally insulating wood laminate. *Chem. Eng. J.* **2020**, *383*, 123109. [CrossRef]
17. Lai, Y.; Liu, X.; Li, Y.; Leonidas, E.; Fisk, C.; Yang, J.; Zhang, Y.; Willmott, J. Investigating the fire-retardant efficiency of intumescent coatings on inclined timber: A study on application strategies and heat transfer mechanisms. *Constr. Build. Mater.* **2023**, *407*, 133586. [CrossRef]
18. Wang, H.Y.; Tian, Y.; Zhang, L. Experimental study of the characteristic parameters of the combustion of the wood of ancient buildings. *J. Fire. Sci.* **2019**, *37*, 117–136. [CrossRef]
19. Qu, W.; Wu, M.; Song, W.; Wu, Y.; Ma, X. Study of a Flame-Retardant Primer Coating Resin Applied on Ancient Timber Structures. *China Wood Ind.* **2020**, *34*, 1–4+12.
20. Roshan, J.P.; Wang, Q.S. Prediction of properties and modeling fire behavior of polyethylene using cone calorimeter. *J. Loss Prev. Proc.* **2016**, *41*, 411–418.
21. Ma, J.; Xiao, C. Large-scale fire spread model for traditional Chinese building communities. *J. Build. Eng.* **2023**, *67*, 105899. [CrossRef]
22. Bei, Y.L.; Yang, D.Y.; Ren, X. Research on the system for fire and personal evacuation of ancient building based on virtual simulation. *Adv. Mater. Res.* **2014**, *1010*, 284–287. [CrossRef]
23. Tian, G.; Chang, K.; Li, A. Research on the development process of ancient building fire based on FDS. *Fire Sci. Technol.* **2019**, *38*, 101–104.
24. Liu, F.; Huai, C.; Li, J. Numerical Analysis of the Development of a Room Fire in a Wooden-structure Historical Building. *ENCE Technol. Eng.* **2019**, *19*, 299–304.
25. Liu, Z. Characterization of indoor fire spread based on FDS. *China Constr. Metal. Struct.* **2021**, *3*, 82–83.
26. Li, X.; Pu, F.; Zou, L.; Gao, G.; Cong, B. Numerical simulation study on influence of plank wall on fire spread in ancient building. *China Saf. Sci. J.* **2019**, *29*, 45–50.
27. Wu, Y.; Hua, B.; Chen, S.; Yang, J. FDS-based study of the fire Properties of Huizhou fire seal walls in traditional residential buildings in southern China. *Fire* **2023**, *6*, 388. [CrossRef]
28. Huai, C.; Xie, J.; Liu, F.; Du, J.; Chow, D.; Liu, J. Experimental and numerical analysis of fire risk in historic Chinese temples: A case in Beijing. *Int. J. Archit. Herit.* **2022**, *16*, 1844–1858. [CrossRef]

29. Zhang, F.; Shi, L.; Liu, S.; Zhang, C.; Xiang, T. The traditional wisdom in fire prevention embodied in the layout of ancient villages: A case study of high chair village in Western Hunan, China. *Buildings* **2022**, *12*, 1885. [CrossRef]
30. Zhao, X.; Wei, S.; Chu, Y.; Wang, N. Numerical Simulation of Fire Suppression in Stilted Wooden Buildings with Fine Water Mist Based on FDS. *Buildings* **2023**, *13*, 207. [CrossRef]
31. Cao, Y.; Chen, X.; Yang, J.; Zhang, H. Heat Analysis and Fire Prevention of Timber Buildings in Southwest China Based on Fractal and Seepage Theory. *Int. J. Heat Technol.* **2021**, *39*, 355. [CrossRef]
32. Li, Y.; Yeo, S.; Dai, S. A Comparative Study of the Fire Properties of Chinese Traditional Timber Structural Components under Different Surface Treatments. *Buildings* **2024**, *14*, 2439. [CrossRef]
33. Bachtiar, E.V.; Kurkowiak, K.; Yan, L.; Kasal, B.; Kolb, T. Thermal stability, fire performance, and mechanical properties of natural fibre fabric-reinforced polymer composites with different fire retardants. *Polymers* **2019**, *11*, 699. [CrossRef]
34. Zhou, B.; Yoshioka, H.; Noguchi, T.; Wang, X.; Lam, C.C. Experimental study on fire performance of weathered cedar. *Int. J. Archit. Herit.* **2019**, *13*, 1195–1208. [CrossRef]

coatings

Article

Essential Oils and Essential Oil-Based Products Compared to Chemical Biocides Against Microbial Patinas on Stone Cultural Heritage

Federica Antonelli [1,2,*], Sara Iovine [3], Carlotta Sacco Perasso [1], Natalia Macro [1], Eleonora Gioventù [4], Filippo Edoardo Capasso [4,5] and Marco Bartolini [1]

1 Biology Laboratory, Istituto Centrale per il Restauro (ICR), Ministry of Culture (MIC), Via di San Michele 25, 00153 Rome, Italy; carlotta.saccoperasso@gmail.com (C.S.P.); marco.bartolini@cultura.gov.it (M.B.)
2 Bio.Co.Ré. Lab, Via Reatina 10, 67068 Scurcola Marsicana, Italy
3 Appia Antica Archaeological Park, Piazza delle Finanze 1, 00185 Rome, Italy; sara.iovine@cultura.gov.it
4 Stone Materials Laboratory, Istituto Centrale per il Restauro (ICR), Ministry of Culture (MIC), Via di San Michele 25, 00153 Rome, Italy; eleonora.gioventu@cultura.gov.it (E.G.); filippoedoardo.capasso@uniroma1.it (F.E.C.)
5 Department of Environmental Biology, Sapienza University of Rome, 00185 Rome, Italy
* Correspondence: fedantonelli@gmail.com

Abstract: In recent years, attention has been paid to finding eco-friendly products that could represent an alternative for the chemicals usually used during conservation procedures. In the field of the devitalization of biological patinas, the most studied products are essential oils (EOs). Several works tested EOs in vitro on microorganisms isolated from stone artifacts but few applied these products on real artworks colonized by a complex patina. In the present work, the biocidal effect of two EOs (cinnamon bark and oregano) and two EO-based products (Biotersus and Essenzio) was compared to that of three chemical biocides commonly used in the conservation of stone artifacts (Biotin R1+R2, NewDes50, and Preventol RI50). The products were applied by brush in situ on a marble slab characterized by a green patina. The biocidal activity was evaluated by analyzing the chlorophyll fluorescence and quantifying ATP through biochemical tests. Furthermore, the products' interference with stone was evaluated by colorimetric measures on sedimentary rock samples. The results indicate that chemical biocides exhibited biocidal activity of greater than 99.5% with a single application, while cinnamon oil (at a 1% concentration) and Essenzio required two applications to achieve the same level of efficacy. Conversely, oregano oil and Biotersus were not able to effectively reduce microorganism vitality, even with repeated applications.

Keywords: cinnamon oil; oregano oil; Biotersus; Essenzio; chlorophyll fluorescence; ATP bioluminescence

1. Introduction

The biological colonization of monuments exposed in outdoor environments involves different groups of microorganisms, both photosynthetic and heterotrophic, and leads to the formation of patinas different in color, shape, and consistency [1,2].

It is well known that the growth of biodeteriogens can lead to significant aesthetic and structural alterations of stone materials due to mechanical and chemical processes. Microorganisms inhabiting stone materials may grow on the surface (epilithic) or in more cryptic habitats such as crevices and fissures (chasmolithic) or may penetrate some millimeters or even centimeters into the rock (endolithic) [3].

The microbial colonization of stone generally starts with a wide variety of pioneer phototrophic microorganisms (mainly cyanobacteria and algae), whose cells are usually immersed in a biofilm rich with organic and inorganic substances and growth factors [1,4]. The deposit of photosynthetic biomass provides an excellent organic nutrient base for the subsequent heterotrophic microflora [2,5].

To avoid the damages produced by biodeteriogens and return stone monuments' legibility, it is of paramount importance to eliminate the biological colonization, preventing future recolonization. Currently, no physical nor mechanical treatments are known to guarantee results comparable to chemical biocides; moreover, in the case of cultural heritage exposed outdoors, the environmental factors cannot be easily adjusted to prevent biological colonization. For these reasons, modern conservation interventions usually require disinfection treatments with biocides. To ensure a high level of protection for human health and the environment, the commercialization and use of biocidal products in the European Union are strictly disciplined by the Biocidal Products Regulation (BPR, Regulation (EU) 528/2012) and the active substances contained in biocides are constantly examined (Delegated Regulation (EU) No. 1062/2014 and Delegated Regulation (EU) No. 825/2022) and removed from the lists of authorized products if considered unsafe or persistent in the environment, causing bioaccumulation phenomena.

Biocides used in the field of cultural heritage conservation, in addition to having to comply with safety criteria for humans and the environment, must possess requirements such as non-interference with the artwork constituent materials and be effective at low concentrations against target organisms. Considering all these limits, the removal from the market of dangerous products in recent decades (e.g., in the case of acutely toxic biocides, suspected teratogenic activity, and high environmental risk) has reduced the number of formulations that can be used for the safe and effective disinfection of stone artifacts.

In recent decades, more attention has been paid in developing conservation procedures safer for conservators and with a reduced impact. In the field of biodeteriogen devitalization, several works were devoted to finding alternative solutions to the chemical biocides currently used for all kinds of materials (such as stone, textile, wood, and paper). Products extracted from bacteria, fungi, and marine organisms have been tested, but the majority of recent experimentations involved the use of plant extracts and, mainly, essential oils and essential oil-based products [6–16].

As defined by ISO, 2013, an essential oil (EO) is the "product obtained from a natural raw material of plant origin, by steam distillation, by mechanical processes from the epicarp of citrus fruits, or by dry distillation, after separation of the aqueous phase—if any—by physical processes".

EOs contain secondary metabolites produced by aromatic plants; they are complex mixtures of hydrophobic and volatile substances such as terpenoids, phenylpropanoids, aliphatic and aromatic alcohols, and aldehydes [17]. Usually, one or two chemical compounds represent most of the percentage weight of the oil (sometimes up to more than 90%) while other components are present at low concentrations or in traces. The chemical processes underlying the antimicrobial-repellent activity of essential oils have not yet been completely understood. It is known that usually the biological effect is linked to the oil main component/components [18]; anyway, studies evidenced that possibly the minor ones in some way influence their activity [19–21].

Being lipophiles, EOs can pass through the cell wall and cytoplasmic membranes and then act through different pathways: altering the polysaccharide layer structure and permeabilizing the membrane, altering proteins and enzyme synthesis, and DNA [17,22–24].

Quantitative and qualitative EO composition is not standard and varies greatly depending on both abiotic (i.e., the environment in which the plant grows, agrotechnical methods, extraction, processing, and storage) and biotic/biological factors (e.g., genetics, parts of the plant, seasonal phase) [25]. When testing EOs' biocidal activity, it is therefore necessary to use oils whose composition and concentration of active ingredients are accurately quantified so that the results obtained can be connected to a precise mixture of substances.

Although EOs are the most tested substances for the devitalization of biodeteriogens, also for those affecting stone monuments, most studies were limited to laboratory trials and tested the oils' biocidal efficacy only on selected microorganisms. Very few works applied these products to complex patinas grown on stone artifacts exposed outdoors [26–34]; so, the topic deserves further investigation.

The present work focused on evaluating the biocidal activity and the possible interference with a sedimentary organogenic rock of two essential oils (cinnamon and oregano), selected based on previous unpublished experimentations carried out in the ICR, and the only two essential oil-based products available on the market (Essenzio and Biotersus), compared to three chemical biocides routinely used in the conservation of stone artifacts exposed outdoors (Biotin R1+R2, NewDes50, and Preventol RI50) [1]. All products were applied twice on the surface, using soft brushes. Even if the application with a poultice would have helped in limiting the evaporation of EOs' volatile components, the aim of the work was to compare the products in conditions that simulate the application in a real conservation intervention. Poultices are in fact used by conservators and restorers for the application of several products (e.g., cleaning operations with inorganic salts); however, for the devitalization of patinas composed of microorganisms and not comprising lichens, especially for large artifacts, the brush application is considered the best choice as it is effective, time-saving, and cost-effective.

2. Materials and Methods

2.1. Products

The study compared the biocidal effect of seven products: Biotin R1+R2 (CTS srl, Rome, Italy), NewDes 50 (CTS srl), Preventol RI50 (LanXess, Cologne, Germany), Biotersus (Exentiae srl, Catania, Italy), Essenzio (Ibix Biocare, Lugo, RA, Italy), cinnamon EO (bark of *Cynnamomun zeylanicum* Blume) (Flora srl, Lorenzana, PI, Italy), and oregano EO (*Origanum vulgare* L.) (Flora srl).

Biotin R1+R2 (from now on BR) is a bicomponent preservative recommended for lapideous materials and organics. The two components, R1 and R2, are composed, respectively, of iodopropynyl butyl carbamate (IPBC) dissolved in diethylene glycol monobutyl ether and n-octyl isothiazolinone (OIT) and terbutryn dissolved in (2-butoxyethoxy) ethanol [35]. The components were mixed in a 1:1 ratio before use and diluted in ethanol, to reach a 3% final concentration. NewDes50 (ND) is a preservative recommended for lapideous materials and organics. It is composed of an aqueous solution of didecyldimethylammonium chloride (DDAC), 50% in concentration [36]. Before application, the product was diluted in deionized water at 5% in concentration. Preventol RI50 (P) is a liquid formulation of alkyl benzyl dimethyl ammonium chloride (benzalkonium chloride) with 50% of an active agent. It is described as a biocide with a broad spectrum of activity covering mold fungi, algae, lichens, bacteria, and yeasts [37]. Before application, the product was diluted in deionized water at 5% in concentration. Both ND and P are biocides based on quaternary ammonium salts; however, the two active molecules have different impacts on the environment, as confirmed by the classification according to CLP regulation reported in the products' safety data sheets (for ND, 'H411: Toxic to aquatic life with long lasting effects'; for P, 'H410: Very toxic to aquatic life with long lasting effects').

Biotersus (B) is a natural product based on an active complex of essential oils for the removal of biological patinas from stone surfaces [38]. The active agent present in the product is the EO of *Thymus capitatus* (L.) Cav., obtained by steam distillation from plant flowers and leaves, containing 50%–75% of carvacrol (information provided by producers). During the experimentation, B was diluted following the ratio suggested by producers (140 mL of B in 10 L of deionized water). Essenzio (E) is a mixture of EOs, among which the most abundant are *Origanum vulgare* and *Thymus vulgaris*. The product is intended to remove mosses, algae, and lichens. It was applied as it is, following the producers' instructions [39].

Cinnamon and oregano EOs from organic farming used during the experimentation were named, respectively, C and O. The chemical characterization of the two EOs was provided by Flora srl. The composition, reported in Table 1, was obtained through gas chromatography–mass spectrometry (GC-MS); instrument, Agilent GC System 7890A

(Agilent, Santa Clara, CA, USA); and column, HP INNOWAX polaire: 60 m, 0.25 mm, and 0.5 μm (Agilent). Two concentrations of the oils were tested: 0.5% and 1%. The solutions were prepared shortly before the application by dissolving the oils in a mixture of water and ethanol, a 30:70 ratio. To evaluate the effect of the water/ethanol mixture (WE) on the microorganisms composing the patina, control tests were performed under the same experimental condition used for the other products.

Table 1. The composition of the EOs tested during the experimentation.

Cinnamon (C)		Oregano (O)	
Component	**%**	**Component**	**%**
α-pinene	3.14	*α*-pinene	0.93
benzaldehyde	0.37	*α*-thujene	1.43
α-phellandrene	0.10	camphene	0.11
α-terpinene	0.10	*β*-pinene	0.11
p-cymene	0.34	*β*-myrcene	1.80
limonene	0.30	*α*-phellandrene	0.15
β-phellandrene	0.25	*α*-terpinene	0.80
linalool	0.25	limonene	0.17
t-cinnamaldehyde	67.70	1.8 cineol + *β*-phellandrene	0.20
cinnamic alcohol	0.18	*γ*-terpinene	5.40
eugenol	5.83	3-octanone	0.20
coumarin	0.25	*p*-cymene	6.32
β-caryophyllene	1.94	terpinolene	0.08
cinnamyl acetate	5.87	1-octene-3-olo	0.44
benzyl benzoate	0.24	*t*-thujanol	0.29
methyl chavicol	<10 ppm	linalool	1.50
		c-thujanol	0.29
		terpinen-4-ol	0.73
		β-caryophyllene	2.64
		α-humulene	0.14
		α-terpineol	0.09
		borneol	0.20
		b-bisabolene	0.15
		thymol	3.34
		carvacrol	70.42
		iso-carvacrol	0.03

Products, the concentration of application, and carriers used are revealed in Table 2.

Table 2. List of tested products, concentration, and carriers used for application.

Product	Acronym	Concentration (%)	Carrier
Biotin R1+R2	BR	3.0	Ethanol
NewDes 50	ND	5.0	Deionized water

Table 2. *Cont.*

Product	Acronym	Concentration (%)	Carrier
Preventol RI50	P	5.0	Deionized water
Biotersus	B	1.4	Deionized water
Essenzio	E	undiluted	-
Cinnamon EO	C (0.5)	0.5	30:70 hydroalcoholic solution
	C (1)	1.0	30:70 hydroalcoholic solution
Oregano EO	O (0.5)	0.5	30:70 hydroalcoholic solution
	O (1)	1.0	30:70 hydroalcoholic solution
Carrier control	WE		30:70 hydroalcoholic solution

2.2. Evaluation of Products' Interaction with Stone

Possible color interference phenomena of the tested products with stone material were evaluated. Tests were carried out in a laboratory on specimens (5 × 5 × 2 cm) of a white sedimentary organogenic rock by means of colorimetric measurements performed before and after products' application. A total of 5 readings per specimen were collected at different randomly selected areas. Standard deviations calculated for each specimen were always lower than 0.5%, so the rock can be considered homogeneous in color.

Biocides were applied on the specimens' surface following the procedure used in the other steps of the experimentation (two applications by brush, three-day interval). Color evaluation was performed after three days from each application using a CM-700d Spectrophotometer (Konica-Minolta, Rome, Italy) with standard illuminant D65 and a 10° observation angle. Color data were expressed according to the CIELAB color system, organized with three axes in a spherical form: L*, a*, and b*. The L* axis is associated with the lightness of the color and ranges from the top (value: 100, white) to bottom (value: 0, black), whereas the a* and b* axes are associated with modifications in redness-greenness (positive a* is red and negative a* is green) and in yellowness-blueness (positive b* is yellow and negative b* is blue).

The ΔE^*_{Lab} after the first and second applications was calculated following the formula for Euclidean distance [40]:

$$\Delta E^*{}_{Lab} = \sqrt{\Delta L^{*2} + \Delta a^{*2} + \Delta b^{*2}}$$

2.3. In Situ Tests

The tests aimed to evaluate the biocidal activity were performed in the Appia Antica Archaeological Park (Rome, Italy), on an archeological marble slab, located next to the Valeri tomb, homogeneously covered by a green patina. Before the start of the experimentation, the patina was sampled with a sterile scalpel and characterized by observing unstained wet mounts through an optical microscope (Axio Imager M2, Zeiss, Oberkochen, Germany) equipped with a digital camera.

Eleven squares, 10 × 10 cm, were delimited, one for each product at the tested concentration, one for WE control, and one for untreated control (NT). The squares were distant from each other by 5 cm to avoid products overlapping due to their diffusion on the surface after application (Figure 1).

Products were applied by brush, to achieve the complete imbibition of the patina. Two applications were performed three days apart. To evaluate the effect of each application, the treated patina was sampled twice, three days after each application. The surface was not washed before sampling in order to evaluate the decrease in cells' viability linked only to the biocidal treatment. The microorganisms were removed from the stone by a sterile scalpel (scraping half of the squares for each sampling), stored in sterile containers, and analyzed on the same day of the sampling. The powder scraped from the stone was weighed and suspended in a sterile saline solution; the final volume of the suspensions was set to obtain the same powder concentration for each sample. Then, suspensions were vortexed for 1 min and aliquots were used to perform the analyses for the evaluation of biocidal efficacy (see Section 2.4).

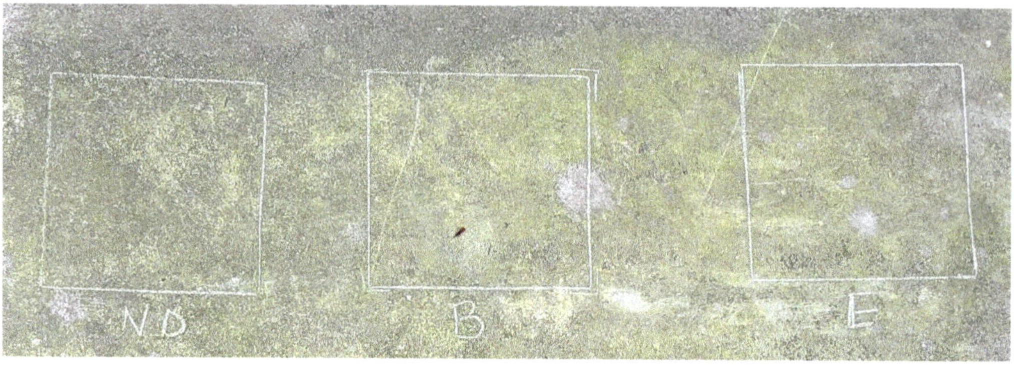

Figure 1. Example of squared areas delimited on the archeological marble slab and used for the application of the tested products.

During the period of application and sampling, no precipitation was reported, and minimum and maximum temperatures ranged between 16 and 24 °C and 28 and 34 °C, respectively. The tested products were always applied in the morning, between 9 and 11 a.m.; during applications, the marble slab was not directly irradiated by the sun. These conditions were selected to prevent a quick evaporation of products after application. Furthermore, it has to be underlined that the marble slab is located in a niche made up of three walls, so it is covered from strong gusts of wind.

2.4. *Evaluation of Biocidal Efficacy*

The biocidal efficacy of the tested products was evaluated by analyzing the chlorophyll fluorescence and quantifying ATP through biochemical tests.

2.4.1. Chlorophyll Fluorescence

The efficacy of biocidal treatments on the photosynthetic microorganisms present in the patina was tested through the evaluation of cell physiological status by chlorophyll autofluorescence [41]. Under blue or UV excitation light, healthy photosynthetic cells show a characteristic red primary fluorescence, caused by chlorophyll a. After a biocidal treatment, cells undergo a gradual pigment degradation and the red color changes from brown to orange and finally to white/greenish, depending on the degree of cell destruction [41].

Tests were carried out on wet mounts obtained from suspensions prepared as reported in Section 2.3; three slides were obtained for each product, using 25 μL of the suspensions. The evaluation of viable cells was conducted by the observation of 10 fields of view identified along the diagonals for each slide. Observations were carried out through an optical microscope (DM750, Leica, Buccinasco, MI, Italy) equipped with a solid-state fluorescence light source (SFL100, Leica), using a 450/490 nm bandpass excitation filter and 515 nm barrier filter.

The vitality level was expressed as the percent decrease in vital cells present in the treated specimens with respect to those in the untreated patina. Based on the values of the percent decrease in fluorescence, four classes of biocidal efficacy were established: class 1 for values lower than 30%; class 2, ranging between 30% and 60%; class 3 for decreases between 60% and 90%; and class 4 for decreases exceeding 90%. Classes comprise wide ranges of values because this technique only provides a qualitative and semi-quantitative evaluation of the treatment effects. In fact, precise cell counting was prevented by the presence of cellular aggregates.

2.4.2. Biochemical Tests

Biochemical tests involved the quantification of the ATP bioluminescence after the two applications of the tested products. The obtained results were compared to the ATP present in the untreated patina. To extract ATP from microorganisms, 20 μL of the suspensions obtained from the sampled patinas (see Section 2.3) was added to 80 μL dimethyl sulphoxide (DMSO) at a concentration of 90% in a TAE buffer (a buffer solution containing a mixture of a trishydroxymethyl-aminomethane base, acetic acid, and ethylene diamine tetra-acetic acid), at pH 7.7. DMSO was preheated in a sterile 1.5 mL Eppendorf tube at 100 °C for two minutes; after adding the suspension, it was incubated for one minute at 100 °C and then rapidly cooled in an ice bath [7,42]. The cellular ATP bioluminescence was evaluated by adding to the lysate 100 μL of a detection reagent from the ENLITEN® ATP Assay System Bioluminescence Detection Kit (Promega, Madison, WI, USA). Blank tests were performed by substituting the sample with the same volume of an extraction reagent. The cellular ATP values reported in the results were obtained subtracting the blank values from the registered cellular ATP. All the tests were performed on three aliquots of the sample, and the results are reported as the average value of the three readings. The bioluminescence was registered as a relative light unit (RLU) through a luminometer LB 9526 Sirius 2 PC-1 (Berthold, Bad Wildbad, Germany) combined with FB12/Sirius Software V2.0 (Berthold) using the following setting parameters: 2 s delay time after reagent injection and 10 s RLU signal integration time. RLUs were converted into pg of ATP by creating a calibration curve with an ATP standard solution.

Results are expressed as a percentage of the ATP decrease with respect to the untreated patina. Based on the obtained results, four classes of efficacy were obtained: class 1, for decreases lower than or equal to 90%; class 2, for values ranging between 90% and 99%; class 3, for decreases ranging between 99% and 99.4%; and class 4, for values exceeding 99.5%. In the data analysis, it should be noted that the difference between a decrease of 99% and 99.9% corresponds to a 10 times greater level of disinfection effectiveness.

3. Results

3.1. Evaluation of Products' Interaction with Stone

Colorimetric results can be seen in Table 3 and Figure 2. Given the higher negative values of ΔL^* after the second application for almost all the products, the data seem to indicate as a general trend the progressive darkening of the surface following the use of the biocides tested in this project. However, the final values are only slightly lower (ΔL^* variation < 1) for all the products and the darkening seems to be significantly less marked (ΔL^* variation < 0.20) for Essenzio, cinnamon essential oil, and oregano essential oil (where the final delta is actually slightly higher after the second application, suggesting that this product should not be responsible for any darkening of the surface). For all tested products, the calculated ΔE^*_{Lab} values increased after the second application. The higher variations between first and second applications were calculated for BR and ND (respectively, 0.99 and 0.81), while for the other products, the values in ΔE^*_{Lab} variation stayed under 0.60. However, for the majority of the products, the calculated ΔE^*_{Lab} was always lower than 2, suggesting that only an experienced user would be able to see any color difference. Furthermore, for oregano essential oil, the final ΔE^*_{Lab} was lower than 1, implying that the color difference was not observable at all to the naked eye [40]. Finally, for several products (Preventol, Biotersus, Essenzio, and cinnamon oil), the $1 < \Delta b^* < 2$ could suggest the possible progressive yellowing of the surface after the application of the biocides.

Table 3. Values of ΔL*, Δa*, Δb*, and ΔE^*_{Lab} after the treatments on sedimentary rock samples.

	Application I				Application II			
Product	**ΔL***	**Δa***	**Δb***	**ΔE^*_{Lab}**	**ΔL***	**Δa***	**Δb***	**ΔE^*_{Lab}**
BR	−0.08	0.00	0.18	0.20	−0.78	0.20	0.87	1.19
ND	−0.11	0.07	0.11	0.17	−0.49	0.09	0.85	0.99
P	−0.20	0.06	1.17	1.19	−0.82	0.16	1.57	1.78
B	−0.36	0.13	0.69	0.79	−0.76	0.16	1.01	1.27
E	−0.59	0.06	1.39	1.51	−0.73	0.09	1.81	1.95
C	−0.35	0.13	1.16	1.22	−0.41	0.08	1.12	1.20
O	−0.27	0.04	0.40	0.48	−0.22	0.07	0.49	0.54
WE	0.11	0.03	0.38	0.40	−0.06	0.03	0.30	0.31

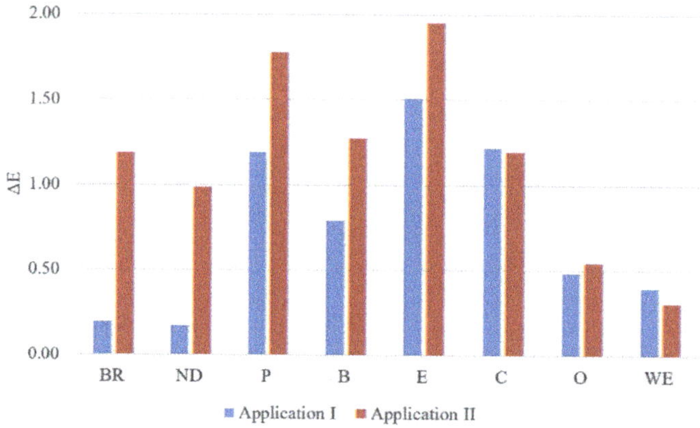

Figure 2. Histogram showing total color variation (ΔE^*_{Lab}) after first and second treatments.

3.2. *In Situ Tests*

3.2.1. Characterization of Patina

The biological patina composition was comparable along the archeological marble slab surface in the monitored delimited squares. The slab was characterized by a thick layer, from dark green to light green in color. The phototropic component was dominant, and it was due to the presence of cyanobacteria with coccal organization referable to the genus *Chroococcus* (Figure 3, black arrow), coccal microalgae belonging to the genera *Pleurococcus* and *Synechococcus* (*S. elongatus*) (Figure 3, blue and green arrows), and subspherical and rod-shaped chlorococcal cells (Figure 3, red arrows).

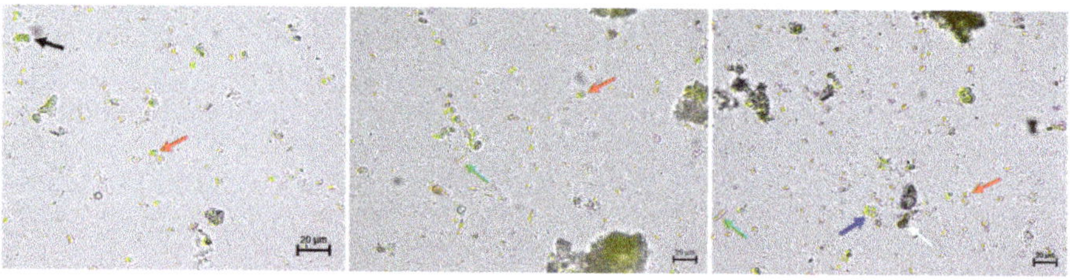

Figure 3. Microbial components of the green patina collected from the archeological marble slab. Microphotographs of coccal microalgae (*Pleurococcus*—blue arrow, and *Synechococcus*—green arrow), subspherical and rod-shaped chlorococcal cells—red arrows, cyanobacteria (*Chroococcus*—black arrow), and microcolonial fungi—white arrow.

The presence of microcolonial fungi (MCF) was also documented (Figure 3, white arrow).

3.2.2. Evaluation of Biocidal Efficacy

Chlorophyll Fluorescence

In Figure 4, three examples of the aspect of photosynthetic cells under visible and UV light are reported. Microorganisms sampled from untreated control (Figure 4a,b) appeared green under visible light and a bright red fluorescence was observed after UV excitation. After two applications of oregano EO, some cells maintained their original shape and color while others appeared to be brown/orange or totally hyaline (Figure 4c). Under UV light, different degrees of fluorescence were observed; the cells that in visible light had preserved the green color were red, while for the others, the red was paler or barely visible (Figure 4d). After two applications of Preventol RI50, cells had completely lost the green color and sometimes appeared as aggregates of amorphous material (Figure 4e). Under UV light, no red fluorescence was observed but only a pale white/greenish halo corresponding to the degraded cells was visible (Figure 4f).

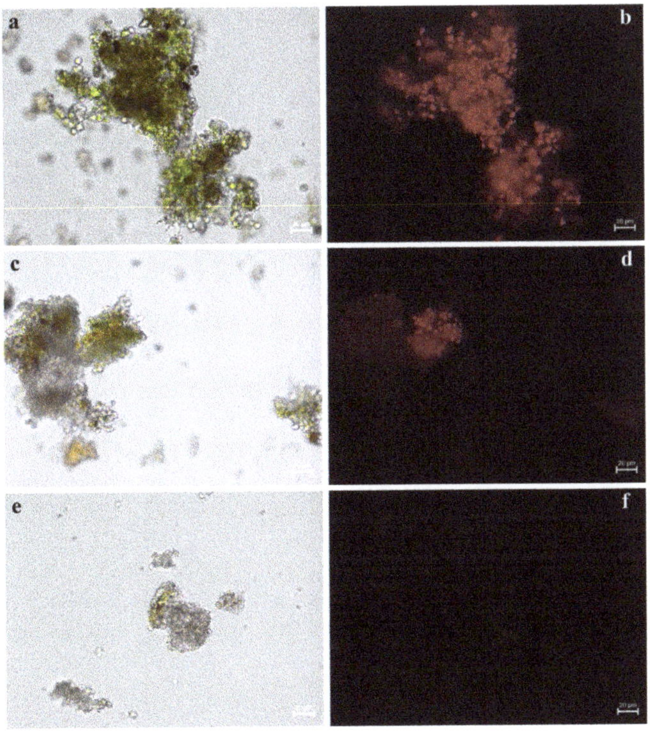

Figure 4. Photosynthetic cells sampled from the archeological marble slab observed under visible (**a**,**c**,**e**) and UV light (**b**,**d**,**f**). (**a**,**b**): untreated control; (**c**,**d**): patina after two applications of oregano EO, 1% in concentration; (**e**,**f**): patina after two applications of Preventol RI50. Figures are intended as an example; the analyses were performed considering 10 fields of view for each analyzed slide.

Table 4 reports the percent decrease in red fluorescent, viable cells after the treatments with respect to the untreated control. After the first application, a percent decrease equal to or over 90% (efficacy class 4) was observed only for P. For ND, E, C (1), and O (0.5), the decrease ranged between 60% and 90% (efficacy class 3), while for C (0.5), O (1), and WE, it was assessed in the range of 30%–60% (efficacy class 2). For BR and B, almost no differences with the NT were observed (efficacy class 1). After the second application, an increase in the values of percent decreases were observed for almost all products. BR, P, E, C (1), and O (0.5) lead to an almost complete disappearance of chlorophyll fluorescence (efficacy class 4). ND, C (0.5), O (1), and WE produced a fluorescence decrease between 60% and 90% (efficacy class 3). Only the results produced by B did not show improvement; in fact, the percent decrease was always lower than 30% (efficacy class 1).

Table 4. The efficacy classification of the tested products after the two applications based on the percentage decrease in photosynthetic cells' fluorescence produced after the treatments with respect to the untreated control.

Product	Efficacy Classes *	
	Application I	Application II
BR	1	4
ND	3	3
P	4	4
B	1	1
E	3	4
C (0.5)	2	3
C (1)	3	4
O (0.5)	3	4
O (1)	2	3
WE	2	3

* Classes of % decrease: 1 = <30%, 2 = 30%–60%, 3 = 60%–90%, 4 = >90%.

The classification of the tested products after the two applications is reported in Table 4.

Biochemical Tests

ATP quantification through biochemical tests achieved a more precise evaluation of the biocidal effect of the tested products. Results reported in Table 5 are expressed as percent decreases in the amount of ATP in the treated samples with respect to NT.

Table 5. Percentage decreases in the amount of ATP in the treated samples with respect to NT and efficacy classifications of the tested products after the two applications.

Product	% Decrease ATP		Efficacy Classes *	
	Application I	Application II	Application I	Application II
BR	99.8 ± 0.0	99.9 ± 0.0	4	4
ND	99.5 ± 0.2	99.9 ± 0.0	4	4
P	99.9 ± 0.0	100.0 ± 0.0	4	4
B	45.4 ± 11.5	89.4 ± 2.5	1	1
E	99.3 ± 0.3	99.5 ± 0.2	3	4
C (0.5)	98.9 ± 0.1	99.1 ± 0.0	2	3
C (1)	98.9 ± 0.3	99.8 ± 0.0	2	4
O (0.5)	97.5 ± 0.5	99.7 ± 0.0	2	4
O (1)	98.7 ± 0.6	98.4 ± 0.9	2	2
WE	98.7 ± 0.7	97.9 ± 0.6	2	2

* Classes of % decrease: 1 = <90%, 2 = 90%–99%, 3 = 99%–99.4%, 4 = 99.5%–100%.

After the first application, P and BR achieved a decrease close to 100% (99.9% and 99.8%, respectively). For ND and E, values comprised between 99% and 99.5%. For all other products, the ATP decrease was lower than 99% and ranged between 98.9% for C and 45.4%, corresponding to B. After the second application, the values of percent ATP decreases improved for almost all products with the only exception of O (1) that remained stable. P, BR, ND, C (1), O (0.5), and E allowed for obtaining values equal or very close (>99.5%) to the total devitalization of microorganisms. C (0.5), O (1), and the control WE ranged between 99.1% and 97.9%. Also, for the second application, the lowest value was obtained for B (89.4%).

The classification of the tested products after the two applications is reported in Table 5.

4. Discussion

The experimentations carried out on stone samples in the laboratory and in situ achieved interesting results concerning the interaction of the tested products with stone material and their efficacy in the devitalization of biological patina. Starting from the first point, P, B, E, and C caused a slight increase in the b value (increase > 1); however, none of the tested products produced a change in the color and aspect of the stone perceivable to the naked eye by a non-experienced observer (all ΔE^*_{Lab} were lower than 2) (Table 3 and Figure 2).

Before discussing the biocidal efficacy, it must be remembered that to protect stone cultural heritage exposed outdoors from biological colonization, it is almost impossible to act indirectly, for example, by modifying environmental parameters that favor the growth of biodeteriogens. Hence, the use of biocides is necessary not only to devitalize microorganisms forming patinas but also to prevent future recolonization. To obtain these results, it is mandatory to achieve the deep disinfection of stone (i.e., a decrease in microorganisms' vitality higher than 99.5%) through the biocidal intervention. For this reason, in the present study, biocides are considered effective only if a similar threshold is reached.

This study proved the chemical biocides to be the most effective with respect to all the other products. In fact, according to biochemical analyses, BR, ND, and P reached the highest class of efficacy after the first application (Table 5, Figure 5). The observation of chlorophyll fluorescence classified only P in class 4 after the first application (Table 4). Anyway, it must be considered that the fluorescence decrease is linked to the pigment degradation that does not occur immediately after treatment. The speed of chlorophyll degradation is mainly linked to the biocide mode of action. Products interacting with the cell membrane and leading to a quick cellular lysis are those that induce a faster chlorophyll degradation. So, if the three days elapsed between the application of the biocide and the analyses were not enough to lead to the destruction of chlorophyll, this could mean that the observations carried out after the first application could have underestimated the real percentage of devitalized cells.

Regarding EOs, satisfactory results in terms of vitality decreases were obtained for C (1) and O (0.5) after the second application, which allowed for reaching the highest efficacy classes for both fluorescence evaluation and biochemical tests (Tables 4 and 5, Figure 5). Increasing the concentration of O to 1% did not lead to better results; on the contrary, the efficacy was slightly lower than O (0.5). A similar effect for oregano EO was observed by Spada and colleagues [26], who observed the best efficacy for 0.75% *w*/*w*, while higher concentrations gave similar or worst results. Other studies tested oregano and/or cinnamon EOs on stone specimens or artifacts [26,27,29–31,34] at different concentrations (0.75%, 1%, 2%), dissolved in a solvent/medium (water, agar–agar, emulsifier, polymer-based nanocapsules) and applied by a brush or poultice, and obtained encouraging results. Anyway, an effective comparison between the cited works and the results obtained in the present study is not easy to achieve mainly because of two factors: the differences in EOs' composition and the different methods used to evaluate efficacy. As said, the EOs' composition is not standard and depends on several factors and the biocidal effect strictly depends on different molecules' presence/absence and concentration (see Section 1); for this reason, a chemical characterization of EOs should always be performed when using these oils as biocides. However, in many published studies, the oils' composition is often not reported or only the chemical specificity (i.e., the most represented molecule and a range of possible percent concentrations) is indicated. For these reasons, it is difficult to discuss and compare the biocidal efficacy of oils with different and unknown compositions. An additional issue preventing the comparison of available datasets is the lack of a standardization in the evaluation of the biocidal efficacy. Many different methods and instruments have been applied for the evaluation of the biocidal efficacy in literature works (e.g., UV imaging, colorimetry, ATP bioluminescence evaluation through hygiene and cleaning evaluation systems). Obviously, the use of such different methods makes it difficult (or even impossible) to compare data.

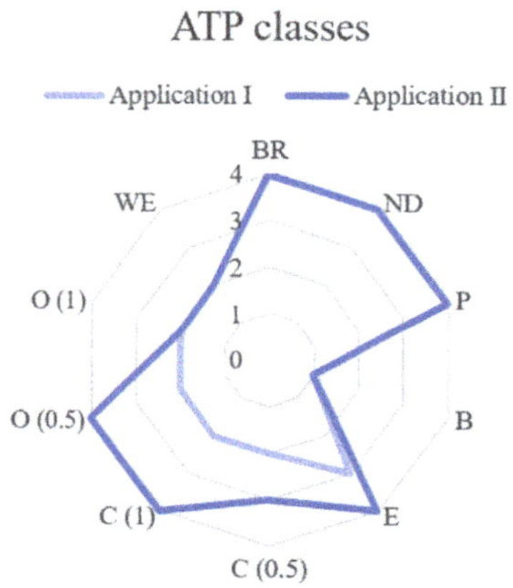

Figure 5. Radar charts reporting the efficacy classification of tested products based on biochemical tests.

It is worth noting that carrier control WE after the first application gave results comparable to those of the two essential oils; after the second application, it was classified in the same class of C (0.5) and O (1) according to fluorescence data, and of O (1) according to ATP results. A similar behavior was observed in tests conducted to evaluate the biocidal activity of essential oils for the protection of waterlogged archeological wood [7]. In that case, the water–ethanol solution had allowed for obtaining ATP percent decreases comparable to those obtained with cinnamon and thyme EOs on two of the fungal strains selected for the in vitro tests (*Chaetomium* sp. and *Stachybotrys chartarum*). Anyway, in that case, cultural tests had proved that the fungi treated with WE were still alive and capable of proliferating on new culture media, so the decrease in the ATP amount was attributed to a fungistatic and not fungicidal activity. In the present study, it is not possible to hypothesize a microbistatic and microbicidal effect since tests of recolonization were not carried out.

For the two EO-based products, the obtained results were very different; E was effective after two applications, while B had poor devitalization results, scoring in class 1 for both ATP quantification and fluorescence evaluation after the second application (Tables 4 and 5, Figure 5). It must be remembered that these products are intended and sold for stone surface cleaning, so their biocidal effect is not guaranteed. Anyway, considering their composition, and, in particular, the presence of EOs, a devitalizing effect could be expected. For this reason, other works tested them against stone biodeteriogens (lichens, algae, and cyanobacteria) [9,29,43]. The products were applied once or twice, by a poultice or brush (with or without a paper covering). Favero-Longo and colleagues and Pascarella and colleagues proved that the effectiveness of Biotersus and Essenzio against lichens can be compared to that of chemical biocides, but the results are influenced by the thalli hydration and application protocol [9,43].

The efficacy of the two EO-based products against a green patina composed of photosynthetic microorganisms was lower with respect to Preventol RI50 but higher than EOs (eucalyptus, basil, clove, pine, thyme, and tea tree) applied following the same protocol, 0.4% in concentration in distilled water [29]. Anyway, it has to be underlined that different methods of efficacy evaluation were applied.

5. Conclusions

In conclusion, the results obtained during the present work showed that chemical biocides remain the best choice in terms of efficacy. In fact, on a green patina composed of cyanobacteria, algae, and fungi, they achieved a satisfactory devitalization level after one application by a brush. Also, some of the EOs and EO-based products reached the highest efficacy class (E, C (1), O (0.5)), but a second application was always necessary. The multianalytical approach used in the evaluation of biocidal efficacy on a complex microbial community allowed us to obtain sound results; furthermore, the use of fully characterized EOs will make it possible to make comparisons with future studies using similar products.

Finally, as previously mentioned, this study was aimed at providing guidance to conservators. For this reason, even if there is a risk that the application by a brush might have influenced the efficacy of EOs and EO-based products (given the volatility of their components), this approach was chosen to simulate the conditions of a real conservation treatment. Application by poultices on large surfaces is in fact not common, and spray application has a poor performance due to the high dispersion of EOs with volatility.

Author Contributions: Conceptualization, M.B. and F.A.; methodology, M.B., F.A., S.I. and C.S.P.; validation, M.B. and F.A.; formal analysis, M.B., F.A. and N.M.; investigation, M.B., F.A., S.I. and C.S.P.; resources, F.A. and C.S.P.; data curation, M.B., F.A., C.S.P. and N.M.; writing—original draft preparation, M.B., F.A., S.I., C.S.P. and N.M.; writing—review and editing, M.B., F.A., C.S.P., S.I., N.M., E.G. and F.E.C.; visualization, M.B., F.A. and C.S.P.; supervision, M.B.; project administration, M.B. and F.A.; funding acquisition, E.G. and M.B. All authors have read and agreed to the published version of the manuscript.

Funding: This research was funded by Project PE 0000020 CHANGES - CUP [B53C22003940006], NRP Mission 4 Component 2 Investment 1.3, Funded by the European Union-NextGenerationEU.

Institutional Review Board Statement: Not applicable.

Informed Consent Statement: Not applicable.

Data Availability Statement: Data are contained within the article.

Acknowledgments: The authors express gratitude to Simone Quilici, for permitting the scientific examination of the stone artifacts in the archeological park.

Conflicts of Interest: The authors declare no conflicts of interest. The funders had no role in the design of the study; in the collection, analyses, or interpretation of data; in the writing of the manuscript; or in the decision to publish the results.

References

1. Pinna, D. *Coping with Biological Growth on Stone Heritage Objects: Methods, Products, Applications, and Perspectives*; Taylor and Francis: Boca Raton, FL, USA, 2017; ISBN 9781771885331.
2. Scheerer, S.; Ortega-Morales, O.; Gaylarde, C. Microbial Deterioration of Stone Monuments—An Updated Overview. *Adv. Appl. Microbiol.* **2009**, *66*, 97–139. [CrossRef] [PubMed]
3. Golubic, S.; Friedmann, E.I.; Schneider, J. The Lithobiontic Ecological Niche, with Special Reference to Microorganisms. *J. Sediment. Res.* **1981**, *51*, 475–478.
4. Tiano, P. Biodegradation of Cultural Heritage: Decay Mechanisms and Control Methods. In Proceedings of the 9th ARIADNE Workshop, Bath, UK, 5–6 November 2002; Institute of Theoretical and Applied Mechanics of the Academy of Sciences of the Czech Republic: Prague, Czech Republic, 2001; pp. 1–37.
5. Ortega-Morales, O.; Hernández-Duque, G.; Borges-Gómez, L.; Guezennec, J. Characterization of Epilithic Microbial Communities Associated with Mayan Stone Monuments in Yucatán, Mexico. *Geomicrobiol. J.* **1999**, *16*, 221–232. [CrossRef]

6. Fidanza, M.R.; Caneva, G. Natural Biocides for the Conservation of Stone Cultural Heritage: A Review. *J. Cult. Herit.* **2019**, *38*, 271–286. [CrossRef]
7. Antonelli, F.; Bartolini, M.; Plissonnier, M.-L.; Esposito, A.; Galotta, G.; Ricci, S.; Davidde Petriaggi, B.; Pedone, C.; Di Giovanni, A.; Piazza, S.; et al. Essential Oils as Alternative Biocides for the Preservation of Waterlogged Archaeological Wood. *Microorganisms* **2020**, *8*, 2015. [CrossRef] [PubMed]
8. Barresi, G.; Di Carlo, E.; Trapani, M.R.; Parisi, M.G.; Chille, C.; Mule, M.F.; Cammarata, M.; Palla, F. Marine Organisms as Source of Bioactive Molecules Applied in Restoration Projects. *Herit. Sci.* **2015**, *3*, 17. [CrossRef]
9. Favero-Longo, S.E.; Laurenzi Tabasso, M.; Brigadeci, F.; Capua, M.C.; Morelli, A.; Pastorello, P.; Sohrabi, M.; Askari Chaverdi, A.; Callieri, P. A First Assessment of the Biocidal Efficacy of Plant Essential Oils against Lichens on Stone Cultural Heritage, and the Importance of Evaluating Suitable Application Protocols. *J. Cult. Herit.* **2022**, *55*, 68–77. [CrossRef]
10. Minotti, D.; Vergari, L.; Proto, M.R.; Barbanti, L.; Garzoli, S.; Bugli, F.; Sanguinetti, M.; Sabatini, L.; Peduzzi, A.; Rosato, R.; et al. Il Silenzio: The First Renaissance Oil Painting on Canvas from the Uffizi Museum Restored with a Safe, Green Antimicrobial Emulsion Based on *Citrus aurantium* var. *amara* Hydrolate and *Cinnamomum zeylanicum* Essential Oil. *J. Fungi* **2022**, *8*, 140. [CrossRef]
11. Argyri, A.A.; Doulgeraki, A.I.; Varla, E.G.; Bikouli, V.C.; Natskoulis, P.I.; Haroutounian, S.A.; Moulas, G.A.; Tassou, C.C.; Chorianopoulos, N.G. Evaluation of Plant Origin Essential Oils as Herbal Biocides for the Protection of Caves Belonging to Natural and Cultural Heritage Sites. *Microorganisms* **2021**, *9*, 1836. [CrossRef]
12. Cirone, M.; Figoli, A.; Galiano, F.; La Russa, M.F.; Macchia, A.; Mancuso, R.; Ricca, M.; Rovella, N.; Taverniti, M.; Ruffolo, S.A. Innovative Methodologies for the Conservation of Cultural Heritage against Biodeterioration: A Review. *Coatings* **2023**, *13*, 1986. [CrossRef]
13. Reale, R.; Medeghini, L.; Botticelli, M. Stealing from Phytotherapy—Heritage Conservation with Essential Oils: A Review, from Remedy to Sustainable Restoration Product. *Sustainability* **2024**, *16*, 5110. [CrossRef]
14. Gabriele, F.; Ranaldi, R.; Bruno, L.; Casieri, C.; Rugnini, L.; Spreti, N. Biodeterioration of Stone Monuments: Studies on the Influence of Bioreceptivity on Cyanobacterial Biofilm Growth and on the Biocidal Efficacy of Essential Oils in Natural Hydrogel. *Sci. Total Environ.* **2023**, *870*, 161901. [CrossRef] [PubMed]
15. Ranaldi, R.; Rugnini, L.; Gabriele, F.; Spreti, N.; Casieri, C.; Di Marco, G.; Gismondi, A.; Bruno, L. Plant Essential Oils Suspended into Hydrogel: Development of an Easy-to-Use Protocol for the Restoration of Stone Cultural Heritage. *Int. Biodeterior. Biodegrad.* **2022**, *172*, 105436. [CrossRef]
16. Bartoli, F.; Corradi, L.; Hosseini, Z.; Privitera, A.; Zuena, M.; Kumbaric, A.; Graziani, V.; Tortora, L.; Sodo, A.; Caneva, G. In Vitro Viability Tests of New Ecofriendly Nanosystems Incorporating Essential Oils for Long-Lasting Conservation of Stone Artworks. *Gels* **2024**, *10*, 132. [CrossRef]
17. Bakkali, F.; Averbeck, S.; Averbeck, D.; Idaomar, M. Biological Effects of Essential Oils—A Review. *Food Chem. Toxicol.* **2008**, *46*, 446–475. [CrossRef]
18. Ipek, E.; Zeytinoglu, H.; Okay, S.; Tuylu, B.A.; Kurkcuoglu, M.; Baser, K.H.C. Genotoxicity and Antigenotoxicity of *Origanum* Oil and Carvacrol Evaluated by Ames Salmonella/Microsomal Test. *Food Chem.* **2005**, *93*, 551–556. [CrossRef]
19. Franzios, G.; Mirotsou, M.; Hatziapostolou, E.; Kral, J.; Scouras, Z.G.; Mavragani-Tsipidou, P. Insecticidal and Genotoxic Activities of Mint Essential Oils. *J. Agric. Food Chem.* **1997**, *45*, 2690–2694. [CrossRef]
20. Santana-Rios, G.; Orner, G.A.; Amantana, A.; Provost, C.; Wu, S.Y.; Dashwood, R.H. Potent Antimutagenic Activity of White Tea in Comparison with Green Tea in the *Salmonella* Assay. *Mutat. Res./Genet. Toxicol. Environ. Mutagen.* **2001**, *495*, 61–74. [CrossRef]
21. Hoet, S.; Stévigny, C.; Hérent, M.F.; Quetin-Leclercq, J. Antitrypanosomal Compounds from the Leaf Essential Oil of *Strychnos spinosa*. *Planta Med.* **2006**, *72*, 480–482. [CrossRef]
22. Nazzaro, F.; Fratianni, F.; De Martino, L.; Coppola, R.; De Feo, V. Effect of Essential Oils on Pathogenic Bacteria. *Pharmaceuticals* **2013**, *6*, 1451–1474. [CrossRef]
23. Reichling, J.; Schnitzler, P.; Suschke, U.; Saller, R. Essential Oils of Aromatic Plants with Antibacterial, Antifungal, Antiviral, and Cytotoxic Properties—An Overview. *Forsch. Komplementarmed.* **2009**, *16*, 79–90. [CrossRef] [PubMed]
24. Saad, N.Y.; Muller, C.D.; Lobstein, A. Major Bioactivities and Mechanism of Action of Essential Oils and Their Components. *Flavour Fragr. J.* **2013**, *28*, 269–279. [CrossRef]
25. Can Baser, K.H.; Buchbauer, G. (Eds.) *Handbook of Essential Oils*; CRC Press: Boca Raton, FL, USA, 2009.
26. Spada, M.; Cuzman, O.A.; Tosini, I.; Galeotti, M.; Sorella, F. Essential Oils Mixtures as an Eco-Friendly Biocidal Solution for a Marble Statue Restoration. *Int. Biodeterior. Biodegrad.* **2021**, *163*, 105280. [CrossRef]
27. Devreux, G.; Santamaria, U.; Morresi, F.; Rodolfo, A.; Barbabietola, N.; Fratini, F.; Reale, R. Fitoconservazione. Trattamenti Alternativi Sulle Opere in Materiale Lapideo Nei Giardini Vaticani. In Proceedings of the XIII Congresso Nazionale IGIIC—Lo Stato dell'Arte, Torino, Italy, 22–24 October 2015; pp. 199–206.
28. Bartolini, M.; Pietrini, A.M. La Disinfezione Delle Patine Biologiche Sui Manufatti Lapidei: Biocidi Chimici e Naturali a Confronto. *Boll. ICR* **2016**, *33*, 40–49.
29. Macchia, A.; Aureli, H.; Prestileo, F.; Ortenzi, F.; Sellathurai, S.; Docci, A.; Cerafogli, E.; Colasanti, I.A.; Ricca, M.; La Russa, M.F. In-Situ Comparative Study of Eucalyptus, Basil, Cloves, Thyme, Pine Tree, and Tea Tree Essential Oil Biocide Efficacy. *Methods Protoc.* **2022**, *5*, 37. [CrossRef]

30. Spada, M.; Sorella, F.; Galeotti, M.; Tosini, I.; Cuzman, O.A. Non-Invasive Technologies to Timely Screen out Different Application Conditions of Essential Oils on Stone. *Int. Biodeterior. Biodegrad.* **2021**, *163*, 105285. [CrossRef]
31. Romano, I.; Granata, G.; Poli, A.; Finore, I.; Napoli, E.; Geraci, C. Inhibition of Bacterial Growth on Marble Stone of 18th Century by Treatment of Nanoencapsulated Essential Oils. *Int. Biodeterior. Biodegrad.* **2020**, *148*, 104909. [CrossRef]
32. Gagliano Candela, R.; Maggi, F.; Lazzara, G.; Rosselli, S.; Bruno, M. The Essential Oil of *Thymbra capitata* and Its Application as A Biocide on Stone and Derived Surfaces. *Plants* **2019**, *8*, 300. [CrossRef]
33. Bruno, L.; Rugnini, L.; Spizzichino, V.; Caneve, L.; Canini, A.; Ellwood, N.T.W. Biodeterioration of Roman Hypogea: The Case Study of the Catacombs of SS. Marcellino and Pietro (Rome, Italy). *Ann. Microbiol.* **2019**, *69*, 1023–1032. [CrossRef]
34. Santo, A.P.; Agostini, B.; Cuzman, O.A.; Michelozzi, M.; Salvatici, T.; Perito, B. Essential Oils to Contrast Biodeterioration of the External Marble of Florence Cathedral. *Sci. Total Environ.* **2023**, *877*, 162913. [CrossRef]
35. CTS srl. *BIOTIN R1 + R2 Nuovo Sistema Preservante a Due Componenti Concentrat i per Materiali Organici e Da Costruzione, Diluibile in Solvente*; CTS srl, 2021.
36. CTS srl. *New Des 50 Preservante Concentrato a Base Di Sali Quaternari d'Ammonio per Materiali Organici e Da Costruzione*; CTS srl, 2013.
37. Lanxess. *Preventol RI80/RI50*; Lanxess: Cologne, Germany, 2011; Volume 6.
38. Exentiae. Biotersus. Available online: https://www.exentiae.it/en/biotersus/ (accessed on 4 November 2024).
39. Ibix Biocare. Essenzio. Available online: https://www.ibixbiocare.it/it/prodotti/essenzio (accessed on 4 November 2024).
40. Mokrzycki, W.S.; Tatol, M. Color Difference ΔE—A Survey. *Mach. Graph. Vis.* **2011**, *20*, 383–411.
41. Bartolini, M.; Pietrini, A.M.; Ricci, S. Application of Fluorescence Microscopy to Evaluate the Vitality of Photoautotrophic Biodeteriogens. In Proceedings of the 8th Euroseminar on Microscopy Applied to Building Materials, Athens, Greece, 4–7 September 2001; Georgali, B., Toumbakari, E.E., Eds.; 2001; pp. 511–516.
42. Rakotonirainy, M.; Héraud, C.; Lavédrine, B. Detection of Viable Fungal Spores Contaminant on Documents and Rapid Control of the Effectiveness of an Ethylene Oxide Disinfection Using ATP Assay. *Luminescence* **2003**, *18*, 113–121. [CrossRef]
43. Pascarella, M.; Fondi, V.; Antonelli, F. Studio Di Prodotti Naturali per La Devitalizzazione Dei Licheni Incrostanti Sulle Pavimentazioni in *Opus Sectile* in Un Contesto Archeologico. In *Intervention on Historic Building Surfaces. What About Innovation?* Driussi, G., Ed.; Edizione Arcadia Ricerche Srl: Bressanone, Italy, 2023; pp. 395–406.

coatings

MDPI

Article

Close-Range Photogrammetry and RTI for 2.5D Documentation of Painted Surfaces: A Tiryns Mural Case Study

Georgios Tsairis [1,*], Athina Georgia Alexopoulou [2], Nicolaos Zacharias [3] and Ioanna Kakoulli [4]

1 Department of Conservation of Works of Art and Antiquities, University of West Attica, 12243 Athens, Greece
2 Laboratory of Conservation—Promotion of Visual Arts, Books and Archival Material (ARTICON Lab), School of Applied Arts and Culture, University of West Attica, 12243 Athens, Greece; athfrt@uniwa.gr
3 Laboratory of Archaeometry, University of the Peloponnese, 24100 Kalamata, Greece; zacharias@uop.gr
4 Materials Science and Engineering, University of California, Los Angeles, CA 90095, USA; kakoulli@ucla.edu
* Correspondence: gtsairis@uniwa.gr or tsairisgeo@gmail.com

Abstract: Painted surfaces, regardless of their substrate, possess unique elements crucial for their study and interpretation. These elements include geometric characteristics, surface texture, brushwork relief, color layer morphology, and preservation state indicators like overpainting, interventions, cracks, and mechanical deformations. Traditional recording methods such as handwritten or digital descriptions, 2D scale drawings, calipers, rulers, tape measures, sketches, tracings, and conventional or technical photography fall short in capturing the three-dimensional detail necessary for comprehensive analysis. To overcome these limitations, this paper proposes the integration of two digital tools, Close-Range Photogrammetry (SfM-MVS) and Reflectance Transformation Imaging (RTI), which have become accessible with the advancement of computing power. While other 3D imaging tools like laser scanners and structured light systems exist and may be preferred for very specialized applications, such as capturing the texture of the surface with sub-millimeter accuracy, SfM-MVS and RTI offer a cost-efficient and highly accurate alternative, with 3D modeling capabilities and advanced pixel color fidelity, essential for documenting the geometric and color details of painted artifacts. The application of these highly promising methods to the mural paintings from the Palace of Tiryns (Nafplion, Greece) demonstrates their potential, providing significant insights for art historians, researchers, conservators, and curators.

Keywords: 2.5D; close-range photogrammetry; SfM-MVS; reflectance transformation imaging; (RTI); heritage documentation; painting conservation

1. Introduction

The study of ancient frescoes presents considerable challenges due to their fragile and sensitive nature, as well as the limitations imposed by their state of preservation and inaccessibility. These factors often restrict direct physical interaction and prevent extensive dimensional measurements, making non-invasive and accurate documentation essential. At the same time, their high cultural and historical value, along with their unique artistic and material characteristics, further emphasize the necessity of their thorough study and long-term preservation.

To address these challenges, 2.5D surface reconstruction has emerged as an indispensable tool in archaeological documentation and conservation. This technique enables the creation of high-fidelity digital replicas that preserve the geometric accuracy, radiometric properties, and surface topography of frescoes. Unlike full three-dimensional (3D) reconstruction, the approach employed here focuses on 2.5D surface reconstruction, capturing the relief of the fresco without generating a full volumetric model. By providing a reliable, measurable, and interactive digital twin of the artwork, 2.5D digital modeling facilitates in-depth study, interpretation, and comparative analysis over time. It also serves as a critical resource for conservators, allowing them to assess deterioration, evaluate past interventions, and guide future conservation strategies with unprecedented precision.

Previous research in the 3D recording and computation of painted surfaces has explored various techniques to overcome these limitations. Methods such as 3D laser scanning and structured light scanning have been applied to capture the intricate surface textures of painted artefacts with sub-millimeter precision [1–3]. Although these techniques have produced highly valuable results, particularly in terms of geometric accuracy and surface detail, they are often associated with significant costs and require specialized equipment, potentially limiting their accessibility and widespread application [4].

In response to these limitations, the present paper proposes the integration of two cost-effective, image-based digital tools—Close-Range Photogrammetry using Structure from Motion-Multi-View Stereo (SfM-MVS) and Reflectance Transformation Imaging (RTI). This approach outlines the practical workflow steps for these techniques. These methods have gained widespread availability in recent decades due to advancements in computing power and the increasing accessibility of personal computers. They have been selected for their ability to generate highly accurate 3D models and photorealistic textures, coupled with their cost-efficiency [5–11]. We propose that these tools become fundamental components of any standard workflow for the scientific study of painted artefacts.

SfM photogrammetry is a passive [12,13] (p. 413–414), image-based documentation technique that allows the derivation of accurate, metric, and semantic information from a series of digital photographic images taken with off-the-shelf digital photographic equipment, but processed in specialist photogrammetry software [14] (p. 15). However, SfM differs fundamentally from conventional photogrammetry, as the geometry of the scene, camera positions, and orientation are solved automatically without the need to specify a priori a network of targets that have 3D positions known in advance. Instead, these are solved simultaneously using a highly redundant, iterative bundle adjustment procedure, based on a database of features automatically extracted from a set of multiple overlapping images [15]. It rigorously turns 2D image data into 3D data (like digital 3D models), establishing the geometric relationship between the acquired images and the scene as surveyed at the time of the imaging event [16] (p. 65). Since it is a passive technique with no emitting light, it relies purely on incident illumination (e.g., the sun or artificial lighting) and, in that way, does not physically harm the object material [17]. However, it is essential to follow appropriate procedures with special care, as the objects in question are archaeological artefacts, which are often sensitive to prolonged or intense light exposure. Even though the term SfM photogrammetry is only part of the process of obtaining 3D models from overlapping pictures, it is also widely used to denote a photogrammetry methodology [18,19].

RTI is a non-invasive/non-contact method that has its roots in the principles of raking illumination ([20] (p. 116), [21] (pp. 127–129)), which has been extensively used in museums and other heritage contexts. A raking light photograph is made by casting light across the surface of a painting at a very low angle, highlighting any surface texture or irregularities, including incisions, impasto, raised or flaking paint, damage, and deformations of the canvas or panel. However, as photographs are usually taken with lights in only one, or perhaps two, positions, the information obtained depends largely on the choice of lighting position; a photograph with lights designed to highlight a particular area may not reveal interesting features in another part of a painting [22] (p. 1). PTM overcomes this drawback by allowing virtual re-lighting of the subject from any direction, and subsequently, the mathematical enhancement of the shape and color attributes of the object's surface reveal information about the topology and reflection of the imaged surface in the form of surface "normal" [23] for each pixel [24–27]. This "normal" information indicates the directional vectors perpendicular to the subject's surface at each location, recorded by the corresponding image pixel. Since each encoded normal corresponds to a point on the object, the whole set provides a complete and accurate "description" of its topography. Consequently, PTMs are 2D images containing true 3D information. This ability to document color and true 3D shape information by using normals is the source of RTI's documentary power [28]. The enhancement functions of RTI reveal surface information that is not readily discernable under direct empirical examination of the physical object. Today's RTI software and related methodologies were developed by an international team of researchers and developers at Cultural Heritage Imaging (CHI) [29].

In this study, SfM-MVS and RTI will be applied to a painted lime plaster fragment from the Palace of Tiryns, room 18 (Small Megaron), dated to the LHIIIB 2 period (late 13th century BC). The fragment, catalogued as MN33460, is housed in the conservation laboratory adjacent to the Archaeological Site of Ancient Tiryns (Nafplion, Peloponnese, Greece) [30,31]. The aim is to produce an accurate 2.5D model and a full-scale ortho-image (1:1 scale), alongside a detailed recording of the surface relief.

Subsequent phases of this research will focus on acquiring comprehensive optical data across various regions of the electromagnetic spectrum, which will be superimposed onto the geometrically accurate digital reference model generated by the SfM-MVS process. This digital substrate will represent a 2.5D model, precisely defined in terms of shape and dimension, free from geometric distortions. The combination of this digital substrate with the complementary data from the RTI application will enable the accurate rendering of the color, texture, surface relief, and pattern of the fresco.

While this research is still ongoing, the present study seeks to demonstrate the potential of these methods and to propose the most effective application procedure to achieve optimal results. Due to the absence of available data from physicochemical analyses of the murals, this paper will not present detailed findings concerning the mural itself. Instead, it will serve as a case study to highlight the proposed methodology for the integration of SfM-MVS and RTI in the documentation and study of painted artefacts.

2. Materials and Methods

In this paper, an integrated methodology is proposed, combining Close-Range Photogrammetry, utilizing Structure-from-Motion (SfM) and Multi-View Stereo (MVS) algorithms, with Reflectance Transformation Imaging (RTI). This dual approach aims to capture and accurately model the three-dimensional geometry of painted surfaces while also gathering high-resolution spatial and color data that are crucial for conservation and analytical purposes. The outlined methodological workflow demonstrates the appropriate application of these techniques to achieve highly detailed and accurate 2.5D models of painted artefacts.

To illustrate the effectiveness of this approach, a Mycenaean wall painting from the Palace of Tiryns, dated to the LHIIIB 2 period (late 13th century BC) [31] (p. 315), [32] (p. xvii) was selected as a case study (According to [32] (p. XVII), "the initial excavations of Mycenaean palaces were conducted many years ago, when the only chronological label available to archaeologists was simply 'Mycenaean'". Carl Blegen's threefold subdivision of the Late Helladic period [33] (35 ff and 120 ff) later allowed scholars to place all extant palatial structures within the LH III phase. Further refinement by Arne Furumark enabled

Blegen to attribute the palace at Pylos specifically to the LH IIIB phase, particularly its middle subphase, shortly after excavation commenced. Currently, there is a scholarly consensus that all remaining Mycenaean palaces date to the LH IIIB period (circa the thirteenth century BCE), a dating convention also adhered to in this study). This fragmented lime plaster painting (see Figure 1a) originally part of the floor decoration within the small megaron of the palace (see Figure 1b) [34] (pp. 178–79, PLATE I), was detached from Room XVIII, "women's hall" [34] (pp. 180–81, PLATE II) (see Figure 1c) [32] (p. 47) (Dr. Wilhelm Dorpfeld notes that Homer [35] (Od. IV. 627; XVII. 169; and XXI. 120) describes the use of simply beaten-in clay for the floors of Mycenaean megarons, a feature also observed in earlier phases at Tiryns. However, Dorpfeld points out that "the palace of Tiryns" exhibits a distinctly different type of flooring, specifically a concrete mixture composed of lime and small pebbles or, in some cases, entirely of lime. Within the rooms, these pebbles are generally absent, resulting in smoother floors. Moreover, in several rooms, the mortar has been fashioned into carpet-like patterns with scratched lines. Such features can still be discerned in the men's hall, its vestibule, and the women's hall [34] (275–76)). It is currently preserved as a portable wall painting fragment; it measures approximately 83.0 cm in height, 89.0 cm in width, and 2.5 cm in thickness [31] (p. 315) and is housed in the conservation laboratory adjacent to the Archaeological Site of Ancient Tiryns (Nafplion, Greece). Given the artefact's fragility and cultural significance, all data acquisition was performed in situ within the conservation facility.

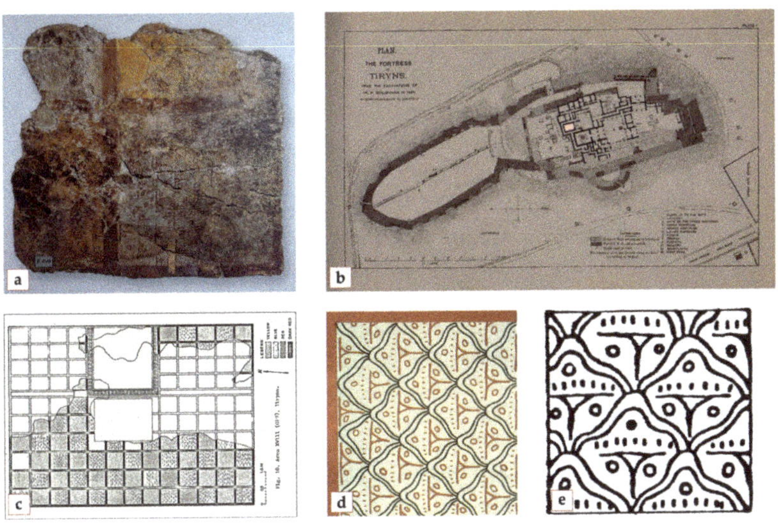

Figure 1. (**a**) The painted floor fragment from room XVIII (Thaler 2018, 315 [31]). (**b**) Plan of Tiryns. Room XVIII is delineated with red parallel line hatching (Schliemann 1885, plate I, modified by G. Tsairis [34]). (**c**) Floor plan of room XVIII. The pattern on the squares is a tricurved arch enclosing a design of lines and circles, identified as flowers on a rock terrain or stone (Hirsch 1974, 47 [32]). (**d**) Reconstruction of the linear patterns from the squares of the Mycenaean floor (Rodenwaldt 1912, 273 [30]). (**e**) A generalized drawing of the linear patterns of tricurved arches (Hirsch 1980, 458 [36]).

The case study artefact exhibits surface irregularities due to the uneven preservation of its fragments, which were uncovered during excavation and consolidated without any attempt to alter their original topography. These preserved variations in surface elevation, along with the artifact's construction technique, make this painting an exemplary candidate for the detailed documentation and analysis afforded by the combined use of SfM photogrammetry and RTI.

In terms of stylistic context, painted floor decorations in the Bronze Age varied significantly between Minoan Crete and the Greek mainland. In Minoan contexts, painted floors in both palatial and non-palatial settings typically featured red pigments on plaster substrates, often arranged in solid color fields, borders for stone slabs, or geometric patterns. Red paint was also used to simulate stone pavements or to create decorative effects within interstitial spaces, occasionally in combination with inlaid materials. Conversely, on the Greek mainland, although Mycenaeans imitated Minoan floor decoration, their decoration was generally limited to the principal rooms of palaces and followed a distinct design approach. The floors were commonly adorned with a grid of intersecting red-painted lines, creating square fields that were colored in blue, yellow, or red, either in isolation or in varied combinations. These fields were further enhanced with linear and marine motifs, often rendered in vibrant colors [36].

Of particular interest in this painted floor fragment is the depiction of tricurved arches, a motif unique to the floors at Tiryns, enclosing a pattern of lines and circles (see Figure 1d) [30] (p. 273); (see Figure 1e) [36] (p. 458). This design portrays a stylized representation of stony or rocky terrain and includes additional elements of "flowers" within each arch as a simplified form of a voluted papyrus blossom in a T shape. The volutes have been reduced to circles on either side of the stem, and the stamen is indicated by a row of dots [37] (p. 229), [32] (p. 48). While the purpose of this floral detail remains unclear—whether it aimed to depict a natural landscape dotted with flowering vegetation or held symbolic significance—it likely functioned as an ornamental enhancement [36].

2.1. Close-Range Photogrammetry (SfM-MVS)

The process involved capturing a series of images from multiple angles, with over 75% overlap, which were then loaded into Agisoft Metashape Pro, a commercial and user-friendly computer vision-based software package, to generate a textured 2.5D model and achieve the ortho-projection of the painting surface. The process is referred to as Structure-from-Motion (SfM) [38] and Multi-View Stereo (MVS) [39] photogrammetry. The majority of studies discussing the application of Structure from Motion (SfM) have focused on the use of Agisoft's Photoscan, later renamed Metashape [40–43]. However, Green et al. [44] highlight the advantages of utilizing open-source software.

SfM is considered an extension of stereo vision. Instead of image pairs, the method attempts to reconstruct depth from several unordered 2D images that depict a static scene or an object from arbitrary viewpoints. It relies on computer vision algorithms [45] that detect and describe local interest points for each image (i.e., image locations that are in a certain way exceptional and are locally surrounded by distinctive texture) and then match those 2D interest points throughout the multiple images. This results in a number of potential correspondences (often called tie points). Using this set of correspondences as input, SfM computes the locations of those interest points in a local coordinate frame and produces a sparse 3D point cloud that represents the geometrical structure of the scene [46]. Inherently, the scale, position, and orientation of any photogrammetric model are arbitrary [47]. For the outputs to be scaled and their dimensions to be measurable, it is necessary to define the relationship between the image and the coordinates of the object. For this purpose, it is essential to place at least three control points (CPs) or GCPs (a GCP is a point on the object illustrated in the image, while at the same time, its 3D coordinates (X, Y, Z) are known, either in a local or in a global reference system) with the exact known distances between them or certified scale-bars established at appropriately selected positions within the photographed scene. The accuracy of the reference information that is used to scale the photogrammetric model will determine the scale accuracy of any data output. After the reconstruction of the 1:1 scale photogrammetric model, any measurement or study of the model can be obtained without limitations regarding the place, as the physical presence in situ, where the original object is located, is not necessary.

SfM photogrammetry as a passive image-based technique, and the results are heavily influenced by the input image data. Employing an automated process to identify and match features by computer vision is fundamentally dependent on the image quality. That being said, any sensors, settings, and acquisition designs should be considered with great care [48].

2.1.1. Photogrammetry: Data Acquisition

The present photogrammetric survey was carried out using a Nikon D850, Full-Frame 45.7 MP Single-Lens Reflex Digital camera with a CMOS sensor (8256 × 5504 pixels, 4.35 μm pixel size), manufactured by Nikon Corporation, Tokyo, Japan, and equipped with a Nikon AF Nikkor 50 mm f/1.8 lens.

In total, 468 images (see Figure 2a–c) in RAW (NEF) format and 16-bit depth were captured by the Nikon D850 camera tethered to a laptop. The mural was placed horizontally on a tray on the floor. The camera was mounted at a distance of 715 mm from the painting, on a custom-made construction of a device [49], allowing fully controlled manual movement along the x and y axes, with horizontal marking to ensure precise movement distance while parallel to the painting surface, achieving constant focus throughout the entire shooting process. A total of 139 images were taken at a small angle of about 15° and −15° to the horizontal plane (see Figure 2b), and 49 images at a larger angle around the mural were taken (see Figure 2c) in order to contribute to the creation of the 3D model keeping constant the focus and distance from the object (Although parallel image acquisitions are good for human stereoscopic projection and automatic surface reconstruction, when combined with convergent acquisitions they often lead to higher accuracy, especially in the z-direction [50] (p. 11), [51] (p. 374)). More specifically, in a coordinate system where the three camera rotation angles are as specified in Figure 3, the images acquired were as follows: 196 images with $\omega = 0°$, $\varphi = 0°$, $\kappa = 0°$; 84 images with $\omega = 0°$, $\varphi = 0°$, $\kappa = 90°$ (see Figure 2a,b); 75 images taken with the camera tilted at an angle of $\varphi = +15°$; and 64 additional images taken with the camera tilted at an angle of $\varphi = -15°$ (see Table 1, Figures 2a–c and 3) to the center of the mural. Two Speedlight flashes in softboxes, to diffuse light evenly, were chosen as a light source, oriented downwards at an angle of 45° to the surface of the wall painting, ensuring uniform illumination and avoiding strong shadows. The use of Speedlight flashes was chosen because they provided adequate lighting conditions, ensuring controlled color balance (white balance), fast shutter speeds of 1/100, a closed aperture of *f*/8 (medium *f*-number), and being in accordance with ISO 100. The two flashes were automatically triggered by the camera using a wireless transmitter at $\frac{1}{4}$ + 0.7 of power.

Table 1. Summarized technical report of the procedure (NIKON D850, Nikon AF Nikkor 50 mm f/1.8 lens, 8256 × 5504 pix, 4.35 × 4.35 μm).

Project	Camera Position	According to Figure 3	Number of Images	Altitude (mm)	Ground Resolution (mm/pix)	Tie Points	Reprojection Error (pix)	F Error (pix)	Scale Bars Error (m)
D850 Tiryns	Parallel, No targets Parallel, targets Turn 90° Tilt 15° Tilt −15° Oblique	$\omega = 0°$, $\varphi = 0°$, $\kappa = 0°$ $\omega = 0°$, $\varphi = 0°$, $\kappa = 0°$ $\omega = 0°$, $\varphi = 0°$, $\kappa = 90°$ $\omega = 0°$, $\varphi = 15°$, $\kappa = 90°$ $\omega = 0°$, $\varphi = -15°$, $\kappa = 90°$	89 107 84 75 64 49	715	0.0555	100,784	1	0.12	0.000013

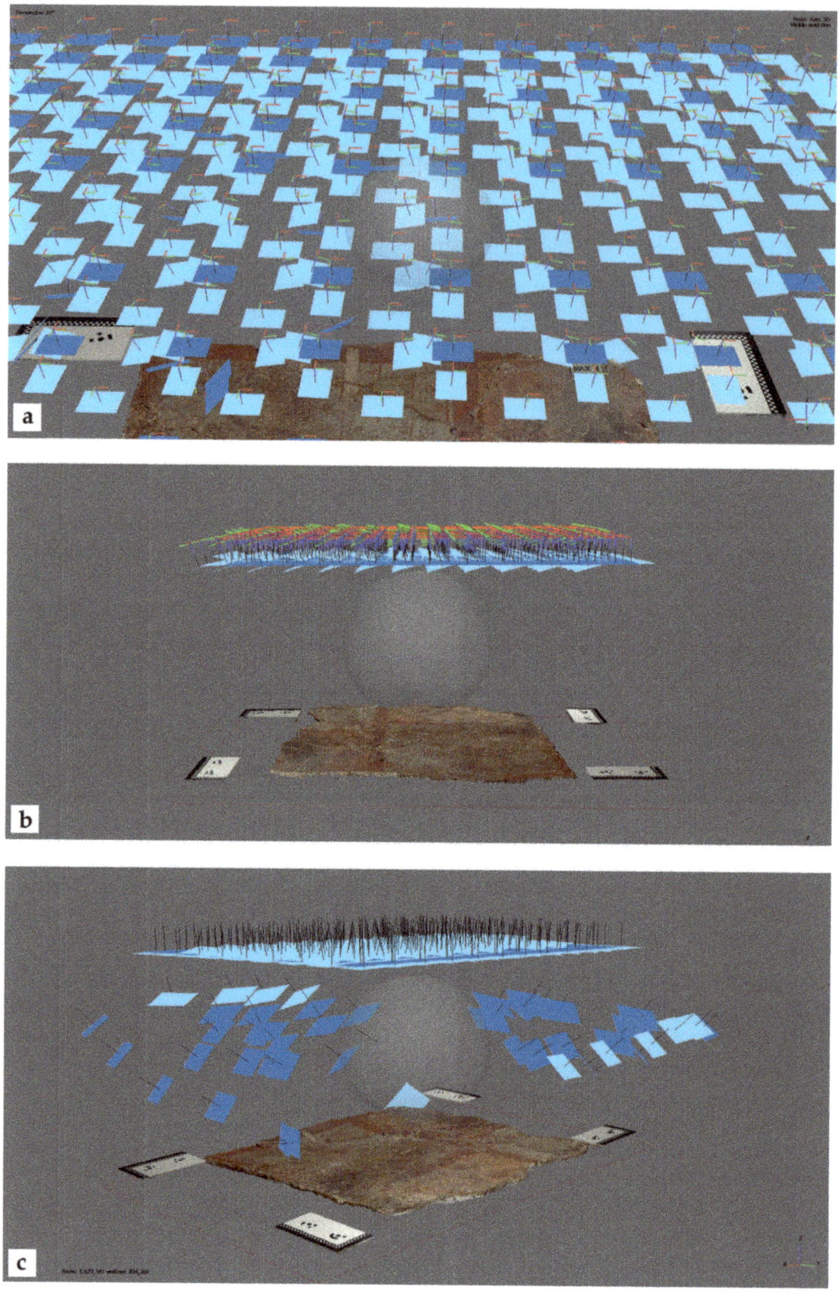

Figure 2. (**a**) The camera positions during the photogrammetric survey are illustrated in blue. A total of 468 images were captured in RAW (NEF) format using the Nikon D850 camera. (**b**) A subset of 139 images was acquired at a slight inclination of 15° and −15° relative to the horizontal plane, which is parallel to the surface of the painting. (**c**) Additionally, 49 images were captured from a steeper inclination around the mural while maintaining a constant focusing distance. The two different shades of blue used to represent the camera positions reflect the significant variation in the camera's inclination during the respective image acquisitions.

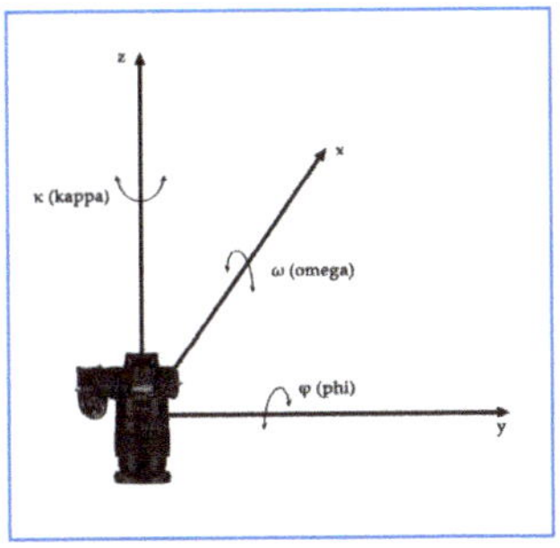

Figure 3. The rotation angles (φ, ω, κ) representing the camera's tilt. The camera movement followed the horizontal plane defined by the x and y coordinates. Illustration by G. Tsairis. Camera image source: Nikon Z6.

Similar to the process of capturing a relatively flat landscape with a UAV (Unmanned Aerial Vehicle) following a flight path designed to ensure sufficient overlap between neighboring strips, a baseline distance of approximately 100 mm was selected along both the x and y axes between successive photographic positions. This systematic approach, analogous to the methodology used in aerial photogrammetry, ensured significant image overlap, thereby enhancing the spatial consistency of the dataset. Specifically, this configuration achieved an estimated 79% overlap in the longitudinal direction, commonly referred to as endlap, and approximately 69% overlap in the lateral direction, which is commonly referred to as sidelap [52] (pp. 718–719), as calculated using the following equations.

The sidelap q was calculated using the following formula:

$$q\% = \left(1 - \frac{SA-A}{SA}\right) \times 100 = \left(1 - \frac{A}{SA}\right) \times 100 = \left(1 - \frac{100\text{ mm}}{318\text{ mm}}\right) \times 100 \Rightarrow \\ q \approx 68.5\%\text{sidelap} \quad (1)$$

where S_A = 318 mm is the lateral image length, and A = 100 mm represents the lateral movement.

Similarly, the endlap p was derived as follows:

$$p\% = \left(1 - \frac{SB-B}{SB}\right) \times 100 = \left(1 - \frac{B}{SB}\right) \times 100 = \left(1 - \frac{100\text{ mm}}{473\text{ mm}}\right) \times 100 \Rightarrow \\ p \approx 78.8\%\text{endlapwhere} \quad (2)$$

where S_B = 473 mm is the longitudinal image length, and B = 100 mm represents the longitudinal movement.

Thus, the configuration yielded a sidelap of approximately 68.5% and an endlap of approximately 78.8%, ensuring substantial image alignment for each camera position. These overlap percentages are critical for minimizing gaps and ensuring comprehensive coverage in both axes, directly supporting the integrity of the dataset for further analysis Figure 4.

To establish the reference system, 16 coded circular targets (calibrated Agisoft Metashape markers) (see Figures 5 and 6) were placed within the subject's frame. Target placement was carefully chosen to avoid areas of high interest on the painting's surface, instead positioning them on a neutral background and at four fixed locations outside the mural. The external targets consisted of custom-made 90° angle rulers, each equipped with prefabricated pairs of Metashape-coded markers separated by highly accurate, pre-measured distances. These rulers were aligned such that the coded targets (5, 1, 7) (see Figures 5 and 6) formed a straight line and were leveled to lie on the same plane with the mural's painted surface.

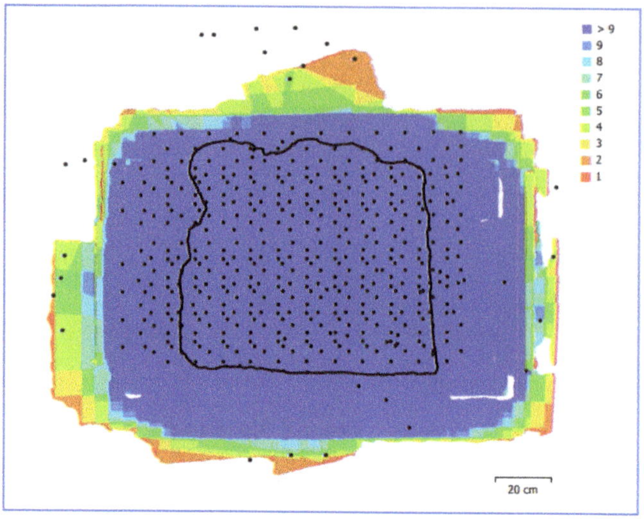

Figure 4. The black dots indicate the positions of the camera, while the different colors represent the number of overlapping images, with the dark blue color (marked as >9) indicating areas where there is overlap from more than nine images. The black outline delineates the surface of the mural. More overlap results in greater accuracy in the estimated measurements.

Figure 5. Screenshot from Agisoft Metashape software, Professional Edition, Version 2.0.1 build 16069 (64-bit). Eight calibrated coded targets (Agisoft markers) were positioned on the surface of the mural. Beyond the mural, four custom-designed scale bars were constructed using 90° angle rulers and four prefabricated coded pairs of Metashape targets (1–8) with precisely pre-measured distances. These scale bars were arranged such that the rulers on each side formed a straight line and were leveled on a plane defined by the coded targets (5, 1, 7) and the mural surface.

By precisely measuring the distance between specific points, referred to as Scale Bars, Agisoft Metashape transforms the spatial data into real-world dimensions, enabling the creation of a 1:1 scale photogrammetric model. During the Align Photo procedure, the software exploits EXIF metadata from the camera and lens to estimate both interior and exterior orientation parameters, including nonlinear radial distortions [53]. Metashape employs Structure from Motion (SfM) and robust feature-matching algorithms, such as Scale-Invariant Feature Transform (SIFT) [54], to automate the estimation of camera geometry without manual calibration. This process incorporates self-calibration, which iteratively refines intrinsic parameters—including focal length, principal point, and lens distortions—

based on correspondences between overlapping images. The bundle adjustment algorithm further optimizes these parameters in conjunction with camera poses and 3D structures, minimizing projection errors and ensuring geometric consistency.

Figure 6. A line laser projecting a linear beam to facilitate the precise alignment of rulers on either side of the mural, ensuring the formation of a straight reference line.

Förstner and Wrobel [52] describe how such methodologies, rooted in photogrammetric principles, enable accurate reconstructions even when using off-the-shelf cameras without pre-calibration. However, while Metashape's automated approach delivers satisfactory results in many cases, projects requiring high precision may benefit from manual calibration to further reduce uncertainties.

In this study, the camera-lens calibration involved capturing 21 images of a checkerboard calibration pattern displayed on an LCD screen, using the same camera, lens, and focal length settings as for the in situ mural imaging. These images were imported into Metashape as a separate chunk, and calibration data were computed and stored. This calibration was incorporated into the photo alignment workflow, enhancing the overall accuracy of the 3D reconstruction process.

In any digital photogrammetric project, the density and precision of measurements are intrinsically tied to three critical variables: the distance from the camera to the object (H), the sensor's pixel size or resolution (ps), and the focal length of the lens (c). These variables collectively determine the ground sample distance (GSD), which represents the spatial resolution of the image and corresponds to the distance between the centers of two adjacent pixels on the subject's surface [47]. GSD is a key metric that influences the accuracy of any photogrammetric analysis. According to [55] (p. 264), the relationship between these variables can be expressed as follows:

$$\frac{c}{H} = \frac{ps}{GSD} \Longrightarrow H = \frac{GSD \times c}{ps} \quad (3)$$

For a camera with a focal length $c = 50$ mm, pixel size $ps = 4.35$ µm, and object distance $H = 715$ mm, the GSD can be calculated as follows:

$$GSD = \frac{H \times ps}{c} = \frac{715\text{ mm} \times 0.00435\text{ mm}}{50\text{ mm}} = \frac{3.11025}{50} = 0.062\text{ mm} \quad (4)$$

Further refinement using Equation (4) with a principal distance of $c = 53.76$ mm yields GSD as follows:

$$GSD = \frac{H \times ps}{c} = \frac{715\text{ mm} \times 0.00435\text{ mm}}{53.76\text{ mm}} = \frac{3.11025}{53.76} = 0.0578\text{ mm}$$

Agisoft software, Professional Edition, Version 2.0.1 build 16069 (64-bit), estimates this value to be 0.055 mm (see Table 1), highlighting that to achieve a GSD below 0.058 mm, images should be captured at distances less than 715 mm from the object. Ref. [56] (p. 31) supports this by relating GSD to the sensor's width and footprint, using the following equation:

$$\frac{p}{GSD} = \frac{f'}{H} = \frac{w}{W} \Longrightarrow GSD = \frac{H}{f'}p \tag{5}$$

where (ps) or (p) represents the photosite pitch, and (H) is the object distance from the optical center. Additionally, based on [57] (p. 46), the principal distance (c) is determined as follows:

$$\frac{1}{f} = \frac{1}{H} + \frac{1}{c} \Longrightarrow c = \frac{H \times f}{H - f} \tag{6}$$

For H = 715 mm and f = 50 mm, this results in a principal distance of c = 53.76 mm. Notably, $c = f$ only when $H \rightarrow \infty$, as explained by [58] (p. 168) and [55] (p. 167).

2.1.2. Photogrammetry: Data Processing

After the sequence of image capture was completed, Adobe Photoshop Lightroom Classic software (12.3 Release) was utilized to adjust the white balance based on the color calibration chart X-Rite Colorchecker® Passport Photo target (in order to relate the recorded colors to the well-defined standards of ICC profiles) and applied to all images via sync. Finally, the RAW files were batch processed into DNG uncompressed format in Adobe Photoshop software (24.4.1 Release) (The term "batch" refers to editing photos of an entire group of images rather than one image at a time, accelerating the editing process).

The next step was to import all images into Agisoft Metashape Pro to generate the 2.5D models (see Table 1). This semi-automated process consisted of seven basic consecutive steps: Align Photos, Build Mesh (3D polygonal model), Build Texture, Build Point Cloud, Build DEM, Build Orthomosaic, and Export Results. A detailed step-by-step guide to the software procedure is presented in the Agisoft Metashape Pro manual. The Build Texture step, during which the color texture map is generated, was performed using 89 images captured parallel to the surface. These images were taken prior to the placement of the coded targets to ensure that the model's surface remained fully visible without being covered by the targets.

2.2. Reflectance Transformation Imaging (RTI) Technique

RTI is a computational photographic method [20] (p. 125) that captures the relief of the object's surface through highlights and shadows in situ, taking advantage of the ability to relight the subject interactively and virtually in real time from various angles at the office. It describes a suite of technologies and methods for generating surface reflectance information using photometric stereo, i.e., by comparison between images with fixed camera and object locations but varying lighting [23]. RTI refers to a file format [5] in addition to a set of methods. The most common implementations of RTI are via Polynomial Texture Mapping (PTM) and Hemispherical Harmonics (HSH) fitting algorithms. PTM, developed by Tom Malzbender and Dan Gelb at HPLabs [24,59,60], is a mathematical model describing luminance information for each pixel in an image in terms of a function representing the direction of incident illumination. The illumination direction function is approximated in the form of a biquadratic polynomial whose six coefficients are stored along with the color information of each pixel [61]. The Hemispherical Harmonics (HSH) algorithm, by contrast, was developed in 2007–2008 by a research team at the University of California, Santa Cruz, under the supervision of Professor James Davis, in collaboration with Cultural Heritage Imaging (CHI), Inc., San Francisco, CA, USA, with consultation provided by Tom Malzbender [62].

Each RTI resembles a single, two-dimensional (2D) photographic image. Unlike a typical photograph, reflectance information is derived from the three-dimensional (3D) shape of the image subject and encoded in the image per pixel, so that the synthesized RTI image "knows" how light will reflect off the subject. When the RTI is opened in RTIViewer software, each constituent pixel can reflect the software's interactive "virtual" light from any position selected by the user. This changing interplay of light and shadow in the image discloses fine details of the subject's 3D surface form [62]. The interactive output is produced from multiple photographs that are taken from one stationary position, while the surface of the subject is illuminated from different raking light positions in each shot. Although this is technically a 2D recording approach, it is often described as 2.5D because of the high-level visual information provided by highlighting and shadowing 3D surfaces. It should be noted that while this procedure provides detailed qualitative surface information, it does not yield metrically accurate 3D data [63].

There are several capture methods that can be utilized to create a Polynomial Texture Map (PTM), each requiring different toolkits and budgets. Due to its relatively low-cost, portable toolkit and flexible recording parameters, the Highlight-RTI (H-RTI) method was selected for this project. H-RTI is a variation of the Reflectance Transformation Imaging (RTI) [5] capture technique, in which the light source is moved manually, and its position is estimated based on reflections detected on a reference sphere placed near the subject. Unlike the classical RTI method, which requires pre-defined and fixed lighting angles, H-RTI offers greater flexibility in data acquisition, making it particularly suitable for in situ applications and the study of objects that cannot be relocated. The H-RTI image capture process allows the recording of light positions as each photograph is taken, obtaining digital image data from which reflectance transformation images (RTIs) can be produced. An RTI not only stores color information for each pixel but also encodes a "normal" value that describes its surface orientation. The processing software computes this value using data from multiple lighting positions relative to the camera, enabling enhanced visualization of surface details.

A key element of the RTI methodology is the presence of one or (preferably) two reflective black glossy spheres in the frame of the photograph in each shot. The reflection of the light source on the spheres, in each image, enables the processing software to calculate the exact light direction for that image. The size of the spheres depends on the dimensions of the object being photographed and the resolution of the camera. It must be noted that the sphere diameter should be at least 250 pixels wide to be used for RTI calculation. As with the various camera setups, the sphere configuration needs to be adjusted according to the circumstances of the subject and environment. The appropriate position for the spheres can be determined by looking at the camera's view.

During post-processing, the reflective spheres will be cropped out from the images; this is something to keep in mind when positioning them. They must be close enough to the subject so that the camera can focus on both the spheres and the subject with sufficient Depth of Field (DoF), but far away enough so that they can be cropped out of the image without losing any image data for the subject itself. It is very important to pay extreme attention during the shooting sequence so that the camera, the target object, and the reflective spheres do not move at all—only the light "moves" [64].

2.2.1. RTI: Data Acquisition

In total, 32 lossless RAW (NEF) format photographs were taken with a Nikon Z6 Full-Frame 24.5 MP mirrorless camera (Nikon Corporation, Tokyo, Japan), equipped with a Tokina AT-X PRO SD 16–28 mm f/2.8 (IF) FX lens (Tokina, Tokyo, Japan), shooting at 28 mm focal length and mounted in a fixed position throughout the process, with the camera sensor parallel to the surface of the painting. The camera was tethered to a laptop via the USB computer-control cable so that it could be remotely triggered via Adobe Lightroom software. Adobe Photoshop Lightroom Classic software (12.3 Release). This layout enabled the camera viewfinder to be remotely displayed in a live-view mode on the larger screen of the laptop, providing the advantage of avoiding any possible movement of the camera while shooting, having better control of the photo frame, focus, brightness, and white balance, as well as checking the quality of the captured images and saving them directly on the laptop's hard drive. A Speedlight flash was employed as the illumination source and was positioned at multiple locations on the surface of an imaginary dome to simulate a three-dimensional lighting arrangement, while the camera remained fixed and stationary throughout the process. Initially, the flash was placed at eight equidistant positions along the perimeter of the dome's base, maintaining a constant distance from the center of the painting surface. These first eight photographs were captured at an elevation angle of 10 degrees above the painting surface. Subsequently, the flash was incrementally elevated along the dome's surface in three more successive steps, each aligned along an imaginary radial line extending from the center of the painting surface toward the dome's surface. These successive positions corresponded to elevation angles increasing in 15-degree increments, reaching from 10 degrees at the base to a maximum elevation of 55 degrees at the highest position. It was automatically triggered by the camera using a wireless transmitter. The use of a flash unit exposes the Mycenaean frescoes to relatively minimal levels of ultraviolet light and allows the image to be captured in 1/100 s at f/8 in ISO 100 (shutter speed 1/100, aperture f/8 and ISO 100).

The quality of the final RTI is based on the quality of the captured images. Although the processing software is relatively robust and photographing in RAW format allows for some post-processing adjustments, it is important to precede test shots with a grey card to select the appropriate exposure (aperture, shutter speed and light intensity from the light source) by examining the histogram of the captured images. This process, at the time of analogue photography, would require the use of a photometer.

It is also recommended that before each capture sequence or during the shooting, a shot be taken using the light source at the highest angle (55 degrees), with a color balance card incorporated in the subject's frame [64]. This image capture will be used later in the post-processing phase to adjust the color of the photographs and ensure accurate color by compensating for the effects of the color temperature of the light source [63]. The color balance card that was used was contained in the X-Rite Colorchecker® Passport [20] (p. 93). Two reflective, black, glossy spheres were positioned within the field of view alongside the mural and remained stationary throughout the acquisition process.

2.2.2. RTI: Data Processing

After the acquisition of the images was complete, Adobe Lightroom software was used to make appropriate white balance adjustments based on the Colorchecker® Passport, which was applied to all images via synchronization. Finally, the RAW files were batch-processed into DNG format and then converted to JPEG (.jpg) in Adobe Photoshop software. Tethered shooting and data processing for both techniques, Close-Range Photogrammetry (SfM-MVS) and Reflectance Transformation Imaging (RTI), was performed by a Microsoft Surface Studio laptop (Microsoft, Redmond, WA, USA) with a Quad-core Intel 11th Gen Intel Core H35 i7-11370H@ 3.30 GHz (Intel Corporation, Santa Clara, CA, USA), 32 GB LPDDR memory (4 × 32 GB), 64-bit operating system (Windows 11, Microsoft, Redmond, WA, USA) and 4 GB GDDR6 GPU memory NVIDIA GeForce RTX 3050 Ti laptop graphics card (NVIDIA Corporation, Santa Clara, CA, USA) was performed by a Microsoft Surface Studio laptop (Microsoft, Redmond, WA, USA) with a Quad-core Intel 11th Gen Intel Core H35 i7-11370H@ 3.30 GHz (Intel Corporation, Santa Clara, CA, USA), 32 GB LPDDR memory (4 × 32 GB), 64-bit operating system (Windows 11) and 4 GB GDDR6 GPU memory NVIDIA GeForce RTX 3050 Ti laptop graphics card.

Post-processing of the 32 image captures occurred through the open-source RTIBuilder software (Version 2.0.2.) and HSH fitter in order to collate information about the direction of the light source and generate the final RTI file. The generated RTI file was then loaded into RTIViewer software (Version 1.1), available from the CHI website [62].

3. Results

3.1. Photogrammetry Results

The outputs generated in the photogrammetric survey of the wall paintings are the Point cloud, the Dense elevation model (DEM), and the Orthomosaics.

The Point cloud consists of points in a three-dimensional space, providing a detailed representation of the geometry of the mural (see Figures 7a,b, 8, and 9).

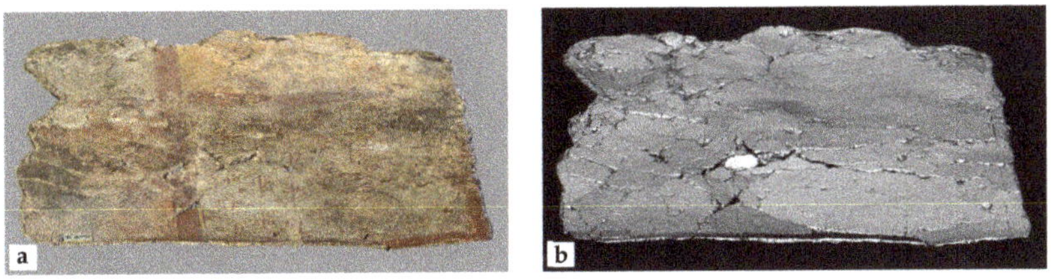

Figure 7. The dense point cloud of the mural, generated through photogrammetric processing and visualized within the Agisoft Metashape software environment, with (**a**) and without (**b**) the application of texture.

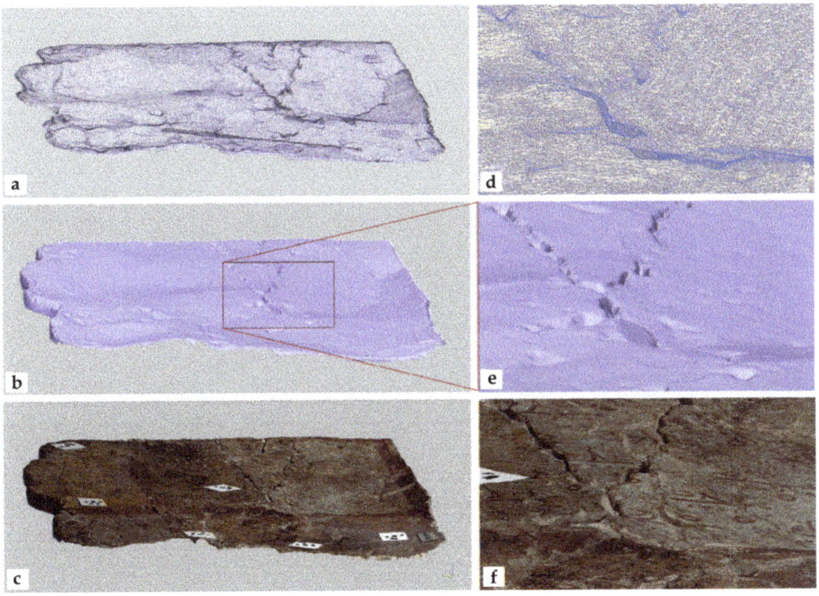

Figure 8. From top to bottom: (**a**) wireframe (triangulated mesh) of the surface, (**b**) 2.5D solid without texture, and (**c**) 2.5D model with texture. (**d**–**f**): Detailed views of (**a**–**c**), respectively.

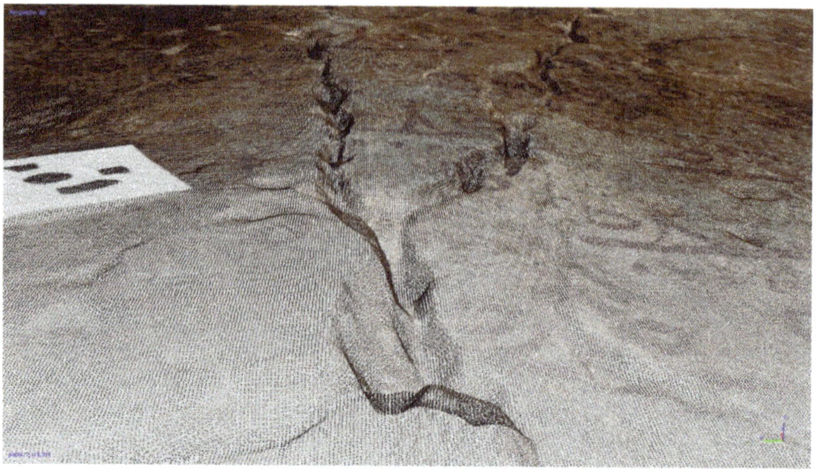

Figure 9. Detailed view of the dense point cloud corresponding to the area depicted in Figure 8d–f. The point cloud consists of points in a three-dimensional space, each carrying color information derived from the source images, providing a high-fidelity representation of the mural's geometry and chromatic characteristics.

The produced Dense Elevation Model (DEM) is a 2.5D model (see Figure 10) of the surface, which uses pixel locations to represent X and Y coordinates and pixel values to represent the depth. The color gradient corresponds to measurable altitudinal gradients. They can also be visualized in the form of elevation contours (see Figure 11). Metashape also enables the calculation of cross sections, with the cut being made at a plane parallel to the z-axis (see Figures 12 and 13).

Figure 10. The generated Dense Elevation Model (DEM) utilizes pixel locations to represent the X and Y coordinates, while pixel values correspond to depth measurements. Additionally, these variations can be represented through elevation contours.

Orthomosaics are usually high-resolution photomosaics (see Figure 14) of numerous overlapping photographs and three-dimensional models (see Figure 7a,b, Figure 8, and Figure 9), providing a metric record of the paintings at a single point in time. The most utilized outputs are orthoimages (see Figures 14 and 15a), which are rectified images that are corrected for most distortions [65].

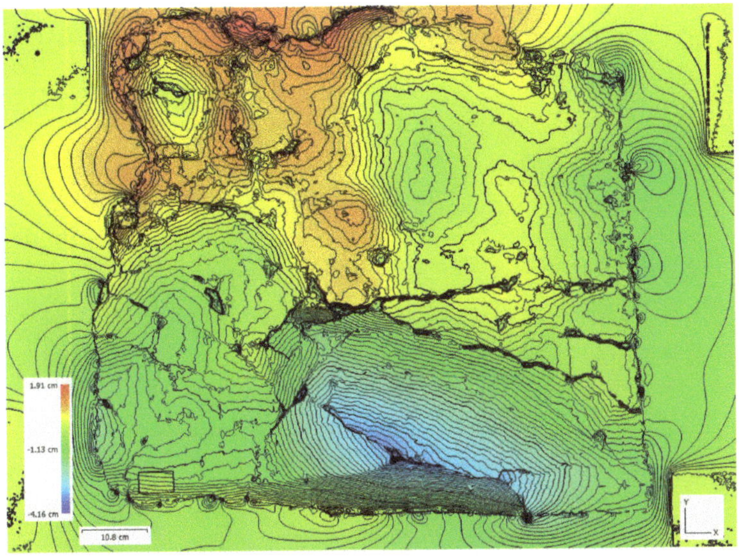

Figure 11. Altitude differences can alternatively be represented as contours. The generated contours are spaced at intervals of 0.001 m.

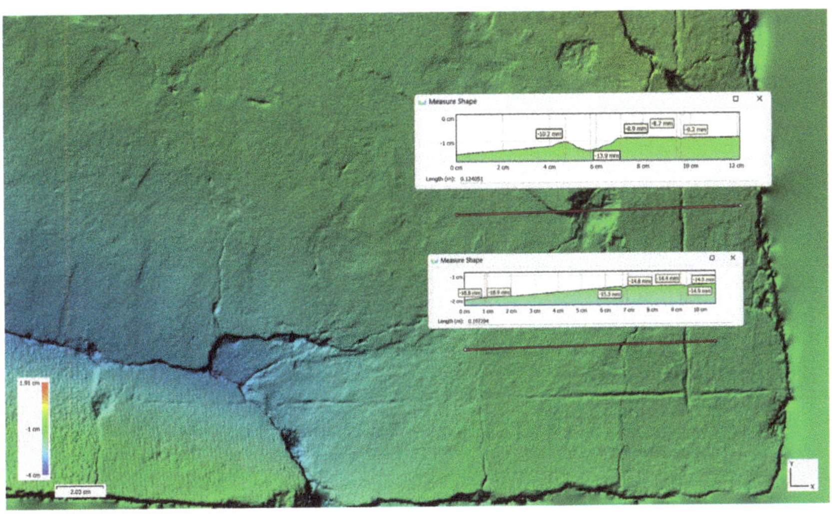

Figure 12. Cross-section providing both visual and metrical data regarding the relief and topography of the mural painting.

The orthoimage is an image of the object surface in orthogonal parallel projection, allowing measurement of distances, angles, areas, etc. (see Figure 15a,b). The orthogonal projection reflects a uniform scale in every product point, since no relief displacements appear on the product. The orthoimage is divided into pixels by considering a specific pixel size based on the ground coordinate system. For each one of these pixels, there is a corresponding specific greyscale or *RGB* value. This value is obtained from the original image corresponding to the appropriate pixel by using one of the resampling methods, namely nearest neighbor, bilinear, and bicubic [55].

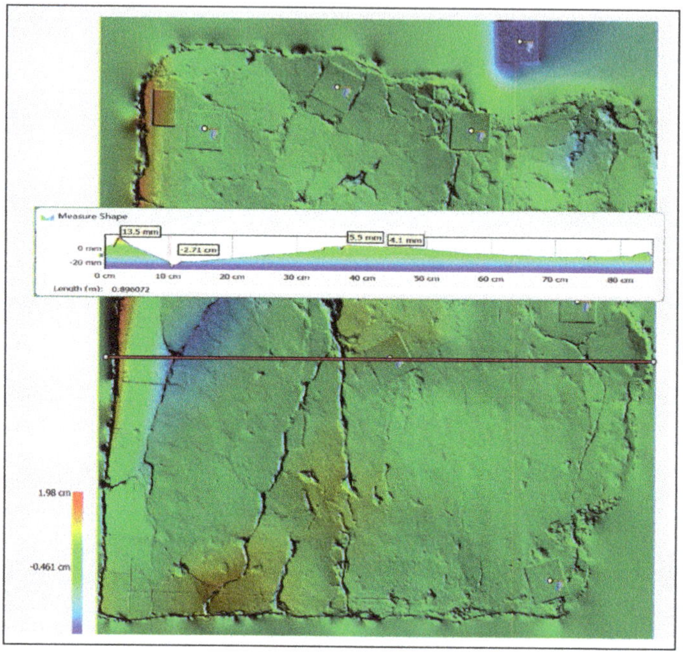

Figure 13. The digital photogrammetry product delivers comprehensive and valuable information pertaining to the object, which may prove useful for future applications. By employing a relatively cost-effective and efficient data collection process, a wealth of information is made available for both current and future use.

Figure 14. The reconstructed orthomosaic generated from 89 DNG images, with an overall resolution of 26,210 × 17,852 pixels and a pixel size of 0.0555 mm. The orthomosaic is orthorectified onto a plane defined by the three targets, 5, 1, and 7 (Figure 5), which are positioned at z = 0 in the local coordinate system.

Figure 15. (**a**) Precise measurements performed on the reconstructed orthomosaic. (**b**) Detailed view of the wall painting, highlighting the ability to obtain accurate spatial data not in situ or on the physical object, but through non-contact methods, virtually accessible from any location, at any time, and on any device.

3.2. RTI Results

The real power of this technique is the interactive RTI Viewer tool (see Figures 16–18), which allows the viewer to virtually relight the object's surface from any direction, much like tilting it back and forth to catch the light and shadow that best reveal features of interest. Furthermore, its enhanced vision reveals shallow reliefs that may not be easily distinguishable under normal lighting conditions or by photogrammetry. However, the results and their interpretation are the subject of an ongoing research project. Therefore, to reach and present confident and reliable conclusions, the information obtained from these two methodologies in combination with other studies should be taken into account. Indicatively, it can be mentioned that the application of the RTI methodology makes it possible to distinguish the painter's brushstroke as well as the sequence of the color layers, i.e., which of them was applied first and which last. That information may be suspected or assumed, but by the RTI, a non-contact, non-invasive technique, it can be certified with certainty.

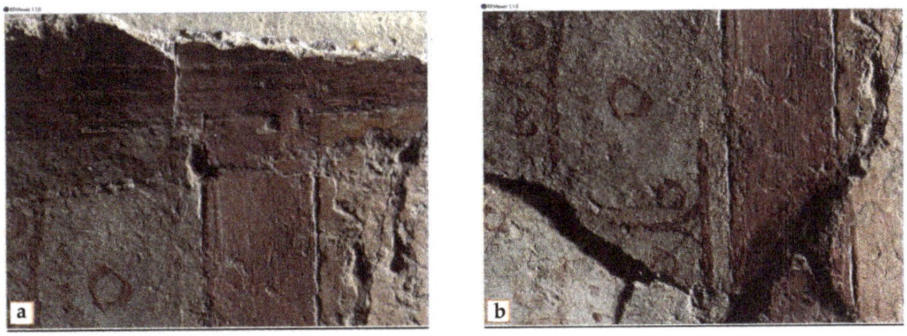

Figure 16. Screenshot from RTIViewer. Reflectance Transformation Imaging (RTI) provides a detailed visualization of surface textures, such as brushstrokes, preliminary incised lines (**a**), and other subtle features that reflect the artist's technique. This method facilitates a deeper analysis of the painting style and an enhanced understanding of the artist's craftsmanship. It also highlights surface anomalies, damages, or irregularities on the painting's surface (**b**).

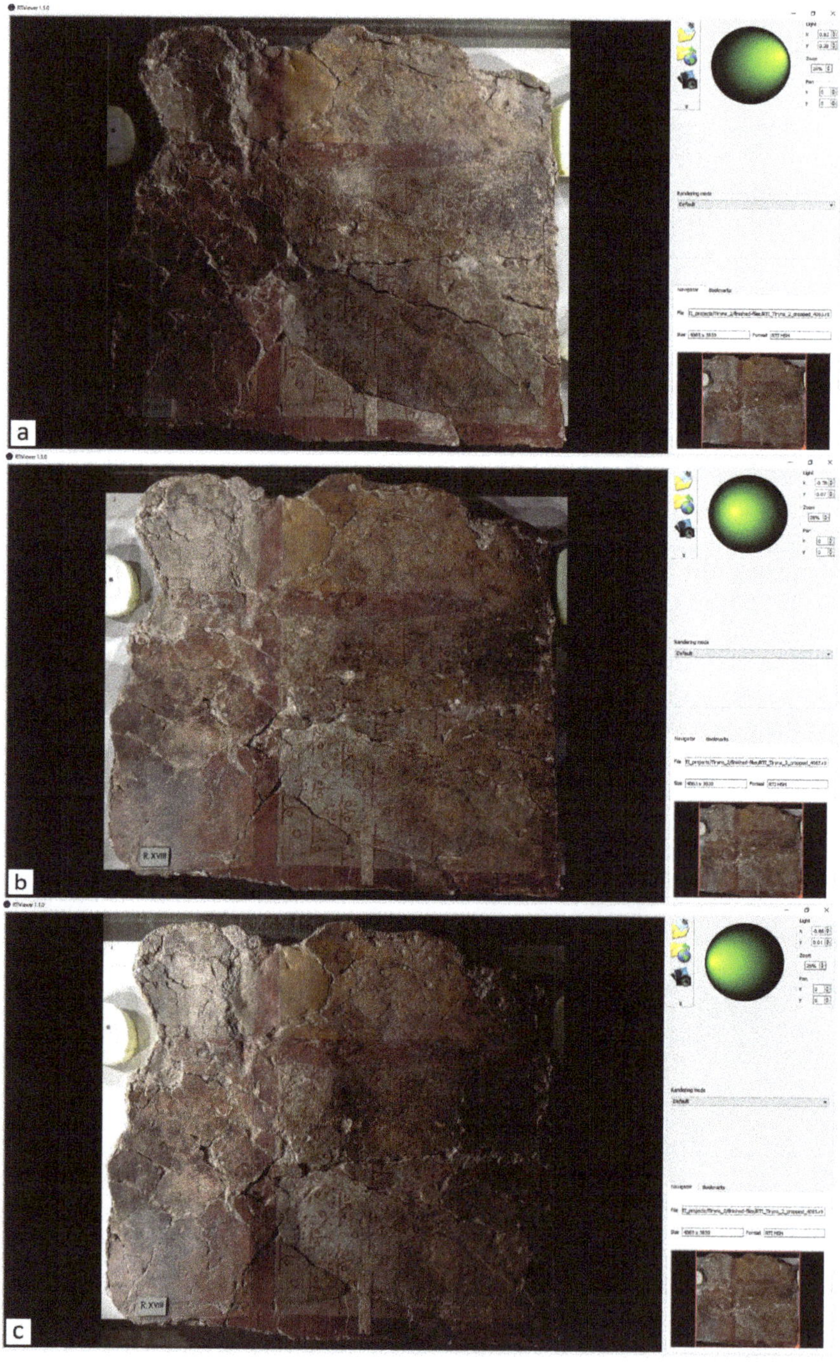

Figure 17. (**a–c**) Screenshots from RTIViewer demonstrating the adjustment of the green sphere (top right) to simulate varying light directions for enhanced surface analysis.

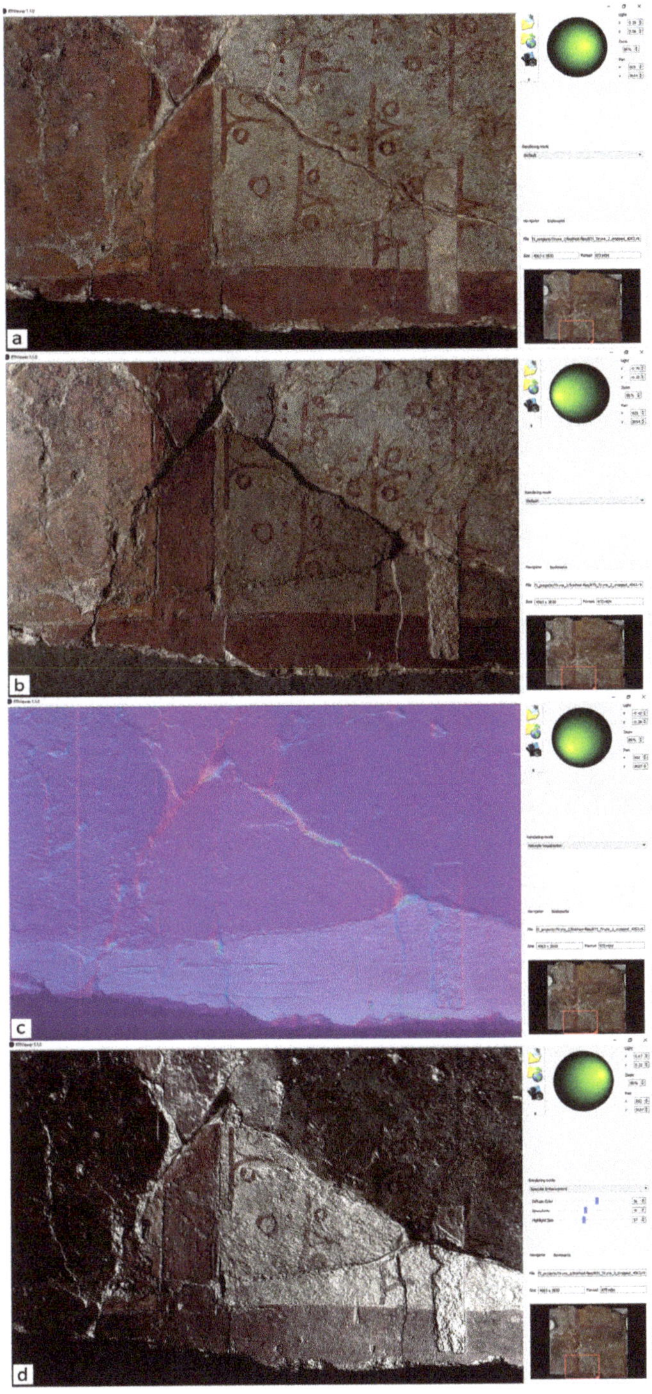

Figure 18. Screenshots from RTIViewer. In (**a**,**b**), different light directions enhance the topography information. In (**c**), the RTI Normals Visualization is utilized, while in (**d**), string impressions are most prominently revealed through the RTI-HSH specular enhancement algorithms.

4. Discussion

This study highlights the invaluable role of Close-Range Photogrammetry (SfM-MVS) and Reflectance Transformation Imaging (RTI) as indispensable tools for the recording and documentation of painted works of art. These methods are proposed as complementary due to the unique and synergistic information they provide. Specifically, photogrammetry generates a precise, geometrically accurate 2.5D digital representation of the artwork's surface, capturing its relief with high fidelity. Meanwhile, RTI, through selective illumination, facilitates virtual relighting and reveals subtle surface texture variations by enhancing them mathematically, thus improving their interpretability.

Moreover, the accessibility of these methods, which rely on consumer-grade cameras, underscores their practical applicability for conservators and researchers, even with minimal technical expertise. However, achieving accurate and reliable results requires strict adherence to the methodology outlined in this study. By following the proposed workflow, practitioners can ensure a robust and standardized approach to the documentation of cultural heritage.

Digitally recording a painting with photogrammetry and RTI and creating the 2.5D model offer the unique ability to exploit depth information, significantly enhancing the study process. This approach overcomes limitations related to the object's preservation status, exposure, storage location, available observation time, or object–observer distance. Regardless of where the physical object is located, its digital "twin" [66] can be accessed and studied from any location, providing the opportunity to examine it closely from a few centimeters away.

A digital replica allows for the observation of an entire scene or a focused area, down to the pixel level, collecting high-resolution spatial and color information. The model can be rotated, zoomed, and measured with absolute precision without physical contact with the original object. Taking advantage of the ability to interactively and virtually relight the subject in real time from various angles in the RTIViewer environment enhances the surface relief, revealing detailed information about construction techniques, color layers, brushstroke textures, and the artist's signature or fingerprint.

These techniques also facilitate the detection of micro-cracks, bulging, flaking, losses, and previous conservation-restoration interventions, which are often not visible to the naked eye. During conservation, they allow for three-dimensional recording of the stages of conservation, damage, or deformation during exhibition or storage, creating a valuable digital record of high geometric and imaging accuracy for future analysis.

This high-accuracy digital record serves as a precisely defined geometric reference for integrating images captured in different wavelengths, drawings, sketches, annotations, and notes. As a geometrically accurate background—free from distortions and with absolute metric precision—it provides a reliable framework upon which all additional data can be corrected and geometrically aligned. This ensures that images from different imaging devices, as well as other forms of documentation, are accurately positioned, enhancing the reliability of study, research, and interpretation. Consequently, this dynamic digital archive strengthens the accuracy and consistency of conservation and scholarly analysis.

In conclusion, the application of these methods is highly promising, and the results justify their proposal as fundamental non-invasive recording tools in the regular workflow of a standardized and integrated scientific methodology for the documentation and study of works of art.

This project is an initial presentation of the methodology proposed for the creation of a geometrically accurate 3D model, a true digital replica, and a full-scale ortho-image (scale 1:1) of the mural under examination as well as the recording of the surface relief and the morphological and geometric characteristics of the color layers and the brushstroke relief utilizing the capabilities provided by Reflectance Transformation Imaging. The resulting orthophoto will serve as the geometrically defined background layer, onto which imaging information from the hyperspectral analysis of the drawing technique and pigments will be superimposed in layers.

Next steps involve acquiring optical 2D information across various regions of the electromagnetic spectrum and overlaying these images, with pixel-level accuracy, onto the digital reference background produced by the SfM-MVS application. This substrate will be a 2.5D digital model, precisely defined in shape and dimensions, free of geometric distortions. Combined with complementary RTI images, this model will accurately render the color, texture, surface relief, and pattern of the painting.

However, research is ongoing to fully exploit these techniques. Due to the current lack of physicochemical study data, a detailed presentation of the mural and an extensive interpretation of the information obtained from the application of SfM-MVS and RTI are not included in this project. Instead, the mural serves as a case study for the application of SfM-MVS and RTI.

5. Conclusions

This study demonstrates the effectiveness of Close-Range Photogrammetry (SfM-MVS) and Reflectance Transformation Imaging (RTI) as complementary, non-invasive tools for the recording and documentation of painted artworks. Their integration provides both geometric and textural information, significantly enhancing the accuracy of digital replicas and conservation workflows. These techniques were applied as a case study on a Mycenaean fresco from the Palace of Tiryns, highlighting their capability to digitally document and analyze fragile cultural heritage artifacts with high precision.

Through photogrammetry, a true digital twin of the original artwork is created—a geometrically precise and metrically accurate 2.5D model that faithfully represents the physical object. This digital replica serves as a reliable foundation onto which additional imaging and analytical data, such as RTI-captured surface details, can be overlaid. The synergy of these techniques results in an enriched dataset that enhances study, conservation, and interpretation.

Beyond its documentation value, this approach is an essential non-destructive and non-invasive method for recording objects with high cultural and historical value of a fragile and sensitive nature. By allowing in-depth examination from a computer screen, it overcomes the limitations of in situ observation, such as restricted access, time constraints, and conservation risks. Researchers and conservators can now analyze fine details with unprecedented precision while ensuring the preservation of the original artifact.

Future work will focus on incorporating optical 2D imaging data across multiple wavelengths, aligning them with pixel-level accuracy onto the digital model. Additionally, further research is needed to integrate physicochemical analysis and enhance the interpretative potential of the recorded data. Despite current limitations, this study highlights the value of SfM-MVS and RTI in developing standardized, high-accuracy digital documentation methods for cultural heritage preservation.

Author Contributions: Methodology, G.T.; Investigation, G.T.; Resources, G.T.; Data curation, G.T.; Writing—original draft, G.T.; Writing—review & editing, G.T. and A.G.A.; Supervision, A.G.A., N.Z. and I.K. All authors have read and agreed to the published version of the manuscript.

Funding: This research did not receive any specific grant from funding agencies in the public, commercial, or not-for-profit sectors.

Institutional Review Board Statement: Not applicable.

Informed Consent Statement: Not applicable.

Data Availability Statement: Data is contained within the article.

Acknowledgments: The authors are grateful to Alkestis Papadimitriou, Director of the Ephorate of Antiquities of Argolida, for the permission to study the floor fragment with painted decoration from the Palace of Tiryns, as well as for providing every assistance at the laboratories of the Ephorate in order to carry out this study. We also thank Aristides Stasinakis and Aristotelis Petrouleas for entrusting us with some of the required photographic equipment and to Metrica S.A. for facilitating the supply of Agisoft Metashape Pro, the commercial software package. Furthermore, the article processing charges (APC) were funded by the Special Account for Research Grants (ELKE) of the University of West Attica.

Conflicts of Interest: The authors declare no conflict of interest.

References

1. Koutsoudis, A.; Vidmar, B.; Ioannakis, G.; Arnaoutoglou, F.; Pavlidis, G.; Chamzas, C. Multi-image 3D reconstruction data evaluation. *J. Cult. Herit.* **2014**, *15*, 73–79.
2. Bianconi, F.; Catalucci, S.; Filippucci, M.; Marsili, R.; Moretti, M.; Rossi, G.; Speranzini, E. Comparison between Two Non-Contact Techniques for Art Digitalization. *J. Phys. Conf. Ser.* **2017**, *882*, 012005.
3. Adamopoulos, E.; Rinaudo, F.; Ardissono, L. Ardissono A critical comparison of 3D digitization techniques for heritage objects. *ISPRS Int. J. Geo-Inf.* **2020**, *10*, 10.
4. Abate, D.; Menna, F.; Remondino, F.; Gattari, M. 3D painting documentation: Evaluation of Conservation Conditions with 3D Imaging and Ranging Techniques. *ISPRS—Int. Arch. Photogramm. Remote Sens. Spat. Inf. Sci.* **2014**, *40*, 1–8.
5. Mudge, M.; Malzbender, T.; Schroer, C.; Lum, M. New Reflection Transformation Imaging Methods for Rock Art and Multiple-Viewpoint Display. In Proceedings of the 7th International Symposium on Virtual Reality, Archaeology and Cultural Heritage, VAST, Nicosia, Cyprus, 30 October–4 November 2006; Volume 6, pp. 195–202.
6. Skarlatos, D.; Kiparissi, S. Comparison of Laser Scanning, Photogrammetry and SFM-MVS Pipeline Applied in Structures and Artificial Surfaces. *ISPRS Ann. Photogramm. Remote Sens. Spat. Inf. Sci.* **2012**, *3*, 299–304.
7. Miles, J.; Pitts, M.; Pagi, H.; Earl, G. Photogrammetry and RTI Survey of Hoa Hakananai'a Easter Island Statue. In *Papers from the 41st Conference on Computer Applications and Quantitative Methods in Archaeology*; Amsterdam University Press: Amsterdam, The Netherlands, 2013; pp. 144–156.
8. Porter, S.T.; Huber, N.; Hoyer, C.; Floss, H. Portable and low-cost solutions to the imaging of Paleolithic art objects: A comparison of photogrammetry and reflectance transformation imaging. *J. Archaeol. Sci. Rep.* **2016**, *10*, 859–863.
9. Altaratz, D.; Caine, M.; Maggen, M. Combining RTI & SFM. A Multi-Faceted Approach to Inscription Analysis. In Proceedings of the Electronic Imaging and the Visual Arts Florence, Florence, Italy, 8–9 May 2019.
10. Kotoula, E.; Robinson, D.W.; Gandy, D.; Jolie, E.A. Computational Photography, 3-D Modeling, and Online Publication of Basketry for Cache Cave, California. *Adv. Archaeol. Pract.* **2019**, *7*, 366–381.
11. Verhoeven, G.; Santner, M.; Trinks, I. From 2D (to 3D) to 2.5 D: Not All Gridded Digital Surfaces are Created Equally. In Proceedings of the 28th CIPA Symposium "Great Learning & Digital Emotion", Beijing, China, 28 August–1 September 2021; Volume 8, pp. 171–178.
12. Kraus, K. *Photogrammetry: Geometry from Images and Laser Scans*, 2nd ed.; Walter de Gruyter: Berlin, Germany, 2007.
13. Remondino, F.; El-Hakim, S. Image-based 3D modelling: A review. *Photogramm. Rec.* **2006**, *21*, 269–291.
14. Kelley, K.; Wood, R.K.L. *Digital Imaging of Artefacts: Developments in Methods and Aims*; Archaeopress Publishing Ltd.: Bicester, UK, 2018.
15. Westoby, M.J.; Brasington, J.; Glasser, N.F.; Hambrey, M.J.; Reynolds, J.M. 'Structure-from-Motion'photogrammetry: A low-cost, effective tool for geoscience applications. *Geomorphology* **2012**, *179*, 300–314.
16. Remondino, F.; Campana, S. *3D Recording and Modelling in Archaeology and Cultural Heritage Theory and Best Practices*; Archaeopress Publishing Ltd.: Bicester, UK, 2014; pp. 65–73.
17. Fuhrmann, S.; Langguth, F.; Goesele, M. MVE-a multi-view reconstruction environment. *GCH* **2014**, *3*, 4.
18. Iglhaut, J.; Cabo, C.; Puliti, S.; Piermattei, L.; O'Connor, J.; Rosette, J. Structure from motion photogrammetry in forestry: A review. *Curr. For. Rep.* **2019**, *5*, 155–168.
19. Solem, D.E.; Nau, E. Two new ways of documenting miniature incisions using a combination of image-based modelling and reflectance transformation imaging. *Remote. Sens.* **2020**, *12*, 1626. [CrossRef]
20. Frey, F.S.; Warda, J.; Heller, D.; Kushel, D.; Vitale, T.; Weaver, G. *The AIC Guide to Digital Photography and Conservation Documentation*; American Institute for Conservation of Historic and Artistic Works: Washington, DC, USA, 2017.
21. Alexopoulou-Agoranou, A.; Chrysoulakis, Y. *Sciences and Artworks*; Gonis, N., Ed.; Gonis: Athens, Greece, 1993.

22. Padfield, J.; Saunders, D.; Malzbender, T. Polynomial texture mapping: A new tool for examining the surface of paintings. *ICOM Comm. Conserv.* **2005**, *1*, 504–510.
23. Woodham, R.J. Photometric method for determining surface orientation from multiple images. *Opt. Eng.* **1980**, *19*, 139–144. [CrossRef]
24. Malzbender, T.; Gelb, D.; Wolters, H. Polynomial texture maps. In Proceedings of the 28th Annual Conference on Computer Graphics and Interactive Techniques, ACM, Los Angeles, CA, USA, 12–17 August 2001; pp. 519–528. [CrossRef]
25. Frank, E. Documenting Archaeological Textiles with Reflectance Transformation Imaging (RTI). *Archaeol. Text. Rev.* **2014**, *56*, 3–13.
26. MacDonald, L.; Robson, S. Polynomial texture mapping and 3d representations. *Int. Arch. Photogramm. Remote Sens. Spat. Inf. Sci.* **2010**, *38*, 422–427.
27. Schädel, M.; Yavorskaya, M.; Beutel, R. The earliest beetle† Coleopsis archaica (Insecta: Coleoptera)–morphological re-evaluation using Reflectance Transformation Imaging (RTI) and phylogenetic assessment. *Arthropod Syst. Phylogeny* **2022**, *80*, 495–510.
28. Happa, J.; Mudge, M.; Debattista, K.; Artusi, A.; Gonçalves, A.; Chalmers, A. Illuminating the past: State of the art. *Virtual Real.* **2010**, *14*, 155–182.
29. Cultural Heritage Imaging (CHI), 2002–2025. [En línea]. Available online: https://culturalheritageimaging.org/ (accessed on 16 January 2022).
30. Rodenwaldt, G.; Hackl, R.; Heaton, N. *Die Fresken des Palastes, 2*; von Zabern, P., Ed.; Eleutheroudakis and Barth: Athens, Greece, 1912.
31. Thaler, U. *Mykene—Die Sagenhafte Welt des Agamemnon*; Wbg Philipp von Zabern in Wissenschaftliche Buchgesellschaft (WBG): Darmstadt, Germany, 2018; p. 391.
32. Hirsch, E.S. Painted and Decorated Floors on the Greek Mainland and Crete in the Bronze Age. 1974. Available online: https://orb.binghamton.edu/dissertation_and_theses/262/ (accessed on 2 October 2022).
33. Blegen, C.W. *Korakou: A Prehistoric Settlement Near Corinth*; American School of Classical Studies at Athens: Boston, MA, USA, 1921.
34. Sturgis, R.; Schliemann, H.; Oxon, D.C.L. *Tiryns: The Prehistoric Palace of the Kings of Tiryns, the Results of the Latest Excavations*; Harper & Brothers: New York, NY, USA, 1886; Volume 2, p. 75.
35. Homerus. *Homēri Odyssea: Scholarum in Usum*; Cauer, P., Ed.; G. Freytag: Lipsiae, Germany, 1887.
36. Hirsch, E.S. Another look at Minoan and Mycenaean interrelationships in floor decoration. *Am. J. Archaeol.* **1980**, *84*, 453–462.
37. Hackl, R. Die Fußboden. In *Tiryns: Die Ergebnisse der Ausgrabungen des Instituts (Band 2): Die Fresken des Palastes*; Rodenwaldt, G., Ed.; Eleutheroudakis and Barth: Athens, Greece, 1912; pp. 222–237.
38. Ullman, S. The interpretation of structure from motion. Proceedings of the Royal Society of London. *Ser. B. Biol. Sci.* **1979**, *203*, 405–426.
39. Furukawa, Y.; Hernández, C. Multi-view stereo: A tutorial. *Found. Trends® Comput. Graph. Vis.* **2015**, *9*, 1–148. [CrossRef]
40. Verhoeven, G. Taking computer vision aloft–archaeological three-dimensional reconstructions from aerial photographs with photoscan. *Archaeol. Prospect.* **2011**, *18*, 67–73. [CrossRef]
41. Verhoeven, G.; Doneus, M.; Briese, C.; Vermeulen, F. Mapping by matching: A computer vision-based approach to fast and accurate georeferencing of archaeological aerial photographs. *J. Archaeol. Sci.* **2012**, *39*, 2060–2070. [CrossRef]
42. De Reu, J.; De Clercq, W.; Sergant, J.; Deconynck, J.; Laloo, P. Orthophoto mapping and digital surface modeling for archaeological excavations an image-based 3D modeling approach. In *2013 Digital Heritage International Congress (DigitalHeritage)*; IEEE: Piscateville, NJ, USA, 2013.
43. Olson, B.R.; A Placchetti, R.; Quartermaine, J.; E Killebrew, A. The Tel Akko Total Archaeology Project (Akko, Israel): Assessing the suitability of multi-scale 3D field recording in archaeology. *J. Field Archaeol.* **2013**, *38*, 244–262. [CrossRef]
44. Green, S.; Bevan, A.; Shapland, M. A comparative assessment of structure from motion methods for archaeological research. *J. Archaeol. Sci.* **2014**, *46*, 173–181. [CrossRef]
45. Szeliski, R. *Computer Vision: Algorithms and Applications*; Springer: Berlin/Heidelberg, Germany, 2022.
46. Verhoeven, G.; Doneus, N.; Doneus, M.; Štuhec, S. From Pixel to Mesh: Accurate and Straightforward 3D Documentation of Cultural Heritage from the Cres/Lošinj Archipelago. In *Istraživanja na otocima*; Hrvatsko arheološko društvo: Lošinjski muzej, Croatia, 2015; Volume 30, pp. 165–176.
47. Barnes, A. Digital Photogrammetry. *Encycl. Archaeol. Sci.* **2018**, 1–4.
48. Georgopoulos, A. Photogrammetric automation: Is it worth? *Mediterr. Archaeol. Archaeom.* **2016**, *16*, 11.
49. Tsairis, G. Development of a Methodology for the Digital Documentation of Fragile Paintings: Utilizing a Low-Cost, Custom-made, Portable Construction for Various Applications in Cultural Heritage Management. In *Book of Abstracts 5th Panhellenic Conference on Cultural Heritage Digitization—EUROMED*; EU: Larissa, Greece, 2024.
50. Linder, W. *Digital Photogrammetry*; Springer: Berlin/Heidelberg, Germany, 2016.
51. Luhmann, T.; Robson, S.; Kyle, S.; Boehm, J. *Close-Range Photogrammetry and 3D Imaging*; Walter de Gruyter GmbH & Co KG: Berlin, Germany, 2023.

52. Förstner, W.; Wrobel, B.P. *Photogrammetric Computer Vision*; Springer: Berlin/Heidelberg, Germany, 2016.
53. Agisoft, L.L.C. *Agisoft Metashape User Manual: Professional Edition, Version 2.0*; Agisoft: St. Petersburg, Russia, 2023.
54. Lowe, D.G. Object recognition from local scale-invariant features. *Proc. Seventh IEEE Int. Conf. Comput. Vis.* **1999**, *2*, 1150–1157.
55. Stylianidis, E.; Remondino, F. *3D Recording, Documentation and Management of Cultural Heritage*; Whittles Publishing: Dunbeath, UK, 2016.
56. Verhoeven, G. Resolving some spatial resolution issues: Part 1: Between line pairs and sampling distance. *AARGnews* **2018**, 25–34.
57. Patias, P.; Karras, G. *Modern Photogrammetric Practices in Architecture and Archaeology Applications*; Diptycho Publications: Thessaloniki, Greece, 1995.
58. Hecht, E. *Optics*, 5th ed.; Pearson: Boston, MA, USA, 2017. Available online: https://emineter.wordpress.com/wp-content/uploads/2020/04/hecht-optics-5ed.pdf (accessed on 9 May 2023).
59. Earl, G.; Beale, G.; Martinez, K.; Pagi, H. Polynomial texture mapping and related imaging technologies for the recording, analysis and presentation of archaeological materials. *Int. Arch. Photogramm. Remote Sens. Spat. Inf. Sci.* **2010**, *38*, 218–223.
60. Malzbender, T.; Gelb, D.; Wolters, H.; Zuckerman, B. Enhancement of Shape Perception by Surface Reflectance Transformation. 2004. Available online: https://www.researchgate.net/publication/220839189_Enhancement_of_Shape_Perception_by_Surface_Reflectance_Transformation (accessed on 3 October 2022).
61. Zányi, E.; Schroer, C.; Mudge, M.; Chalmers, A. Lighting and Byzantine Glass Tesserae. In Proceedings of the EVA London Conference, London, UK, 11–13 July 2007.
62. Cultural Heritage Imaging Reflectance Transformation Imaging (RTI). 2002–2025. [En línea]. Available online: https://culturalheritageimaging.org/Technologies/RTI/ (accessed on 16 January 2022).
63. England, H. *Multi-Light Imaging for Cultural Heritage*; Historic England: Swindon, UK, 2018.
64. *Cultural Heritage Imaging Reflectance Transformation Imaging: Guide to Highlight Image Capture v2.0*; CulturalHeritage Imaging: San Francisco, CA, USA, 2013.
65. Granshaw, S.I. Photogrammetric terminology: Fourth edition. *Photogramm. Rec.* **2020**, *35*, 143–288. [CrossRef]
66. Grieves, M. Digital Twin: Manufacturing Excellence Through Virtual Factory Replication. *White Paper* **2014**, *1*, 1–7.

MDPI AG
Grosspeteranlage 5
4052 Basel
Switzerland
Tel.: +41 61 683 77 34

Coatings Editorial Office
E-mail: coatings@mdpi.com
www.mdpi.com/journal/coatings

www.ingramcontent.com/pod-product-compliance
Lightning Source LLC
Chambersburg PA
CBHW061557210926
51850CB00009B/2129
* 9 7 8 3 7 2 5 8 5 7 3 7 1 *